SS HELMETS

SS HELMETS

THE HISTORY, USE AND DECORATION OF THE HELMETS OF THE BLACK CORPS

Michael D. Beaver *with* Kelly Hicks

Schiffer Military History
Atglen, PA

Disclaimer

This book is about history, and is extremely well suited to militaria collectors, reenactors and other history buffs. Any attempt to interpret the intent of this book as anything other than an historical reference (for example: politics, to glorify evil, Nazism, etc.) is wrong.

Acknowledgements

To those who have invested so much of their time, energy and personal treasure toward this pursuit; all the while showing passion and integrity, this book is dedicated. The following list is by no means all-inclusive and represents only a small fraction of the notable members of the worldwide military collecting community. Special thanks go to those below who have contributed items or period photographs for this book: Al Barrows, Barry Smith, Ben Overhand, Bill Rizzuti, Bill Shea, Bob Biondi, Bob Hoffman, Bob Iqbal, Bob Seijas, Brian Hoesel, Brian Maederer, Bruce Hermann, Byron Vickery, Chris Armold, Chris Jones, Darryl Pajot, David Azar, David May, Derek Chapman, Don Ginsburg, Don Troiani, Ed Hicks, Ed Sayre, Edward Hicks, Frank O'Connor, George Petersen, Greg Aloisio, Gregory Currie, Gus Villarreal, Jim Gauthier, Jim Gray, Jim Toncar, Joe Godfrey, Peter Koeleman, John Atno, John Faracco, John Hale, John Suprin, Jon Moran, Jules Lichenstine, Karl Kithier, Ken Bolton, Ken Bowra, Ken Cavallo, Ken Niewiarowicz, Larry Foley, Larry Marcella, Leighton Longhi, Ludwig Baer, Mark Yerger, Matt Miller, Mika Hankpää, Marc Negrault, Peter Schubert, Mike Davis, Mike Gaspar, Mike Horetsky, Miklos Garamvari, Morten Balestrand, Oswald K. Fernschild, Paul Sanders, Penn Templeman, Perry Floyd, Peter Baert, Peter Whamond, Phil Miller, Philippe Letontourier, Ray Embree, Rex Reddick, Rich Rygaard, Robert Kondor, Robert Moses, Robert Noss, Roger Bender, Ruland Hardy, Russ Hamilton, Shuichi Kamisada, Steve Drucker, Steve Wolfe, Tom Kibler, Tom Murray, Tom Spier , Ty Smith, Willi Schumacher, Willi Zahn, William Matchett.

Book design by Robert Biondi.

Library of Congress Catalog Number: 2005938220.

Printed in China.
ISBN: 0-7643-2476-4

We are always looking for people to write books on new and related subjects. If you have an idea for a book, please contact us at the address below.

Published by Schiffer Publishing Ltd.
4880 Lower Valley Road
Atglen, PA 19310
Phone: (610) 593-1777
FAX: (610) 593-2002
E-mail: Info@schifferbooks.com.
Visit our web site at: www.schiffermilitary.com
Please write for a free catalog.
This book may be purchased from the publisher.
Please include $3.95 postage.
Try your bookstore first.

In Europe, Schiffer books are distributed by:
Bushwood Books
6 Marksbury Ave.
Kew Gardens
Surrey TW9 4JF, England
Phone: 44 (0)20 8392-8585
FAX: 44 (0)20 8392-9876
E-mail: Info@bushwoodbooks.co.uk.
Visit our website at: www.bushwoodbooks.co.uk.
Free postage in the UK. Europe: air mail at cost.
Try your bookstore first.

CONTENTS

INTRODUCTION

Mike Beaver dedicated himself to producing a comprehensive and unequaled work on the uniforms and headgear of the SS. Mike cast his net widely among the best-known collectors internationally, and has come up with a truly representative grouping of the known SS helmet variations, both prewar and wartime. Sadly, before this work could come to fruition, Mike passed away. Prior to his illness, Mike had asked me to team up with him to create captions and text that would put some order and structure to the book. I took on this important task with pride and determination, finishing it in early 2003. In our last conversation a week before his passing, Mike was as engaged as ever in the process. He asked me to be sure to update the text to reflect any important changes that would make the book the ultimate reference on SS helmets. Characteristically, he was providing detailed input as to how he wanted certain things to be sequenced and other aspects – as if he was going to remain engaged in the process for some time to come. That is a testament to Mike's personal commitment as well as the degree of dedication that has gone into the production of this book.

• • •

It is worth noting that we still know relatively little about helmets used by the SS, even sixty years after their demise. Little documentary information exists regarding the manufacture, quality control, and distribution aspects of their helmets. We do know, for example, that the SS were given their own helmet in 1934: the SS-RZM helmet. This was a mass-produced and widely distributed special contract helmet. But on the other hand, we do not know what the simple "batch" numbers inside the military-procured helmets actually refer to – this will be discussed in more detail later in this book. But there are other mysteries about SS helmets. For example, why did the SS wear double runes on their helmets? Why did they reverse their decals frequently? These types of anomalies simply do not exist in the other services. Hopefully the presentation of the materials in this book will shed some light on the lesser understood aspects of SS helmets.

Kelly Hicks, Lieutenant Colonel (Ret.)
New Jersey
July 2005

CHAPTER ONE

DAS SCHWARTZE KORPS

"The black helmet is particularly impressive. It gives us a martial air, very menacing. To tell the truth we also look slightly sinister." — Peter Neumann, The Black March, New York 1960, p.88.

Virtually all of the SS units used a mixture of helmet shell styles and patterns throughout the early Nazi period, seemingly without trying to adhere to one style of helmet shell, but generally using all available types. Within one unit can be observed a variety of shells in use in the same photograph. So given the scarcity of resources available to equip the SS in the early 1930s, units used whatever they could find or contract for on a limited basis. Contracting, such as the SS-RZM helmet, will be discussed in more detail in subsequent chapters.

1934 appears to have been the year that the SS began to use formally accepted methods of helmet decoration. We will examine this practice chronologically, with period photographs and photographs of existing original examples.

Manufacturers of the various helmets in use by the SS in 1934 include the manufacturers from World War I Germany and the Austria-Hungary Empire; as well as commercial manufacturers of early 1930s Germany. Some of these helmets are readily identifiable in the abundant photographic record as German M-16, M-17 and M-18 shells; Austrian M-16 shells, "mock Austrian M-16" shells, "droop bills," SS-RZM helmets, "Himmler helmets," several civic varieties as well as Vulkanfiber helmets. Firms such as Eisenhuttenwerke-Thale (ET), Berndorfer (EB-Austria), Quist and others manufactured the World War I shells. Others, such as the civic and lightweight shells, were manufactured by a sizeable variety of smaller commercial firms such as Erel, Edelstahl, Thalestahl, etc.

A brief word on manufacturer characteristics is in order. Commercially produced helmets generally show the following features as outlined below:

Maker: Unknown - single
Nomenclature : "SS-RZM"
Year of manufacture: 1934
Shell characteristics: Steel-with flow around skirt midway down. Issued in black and possibly anodized finishes.
Liner: Factory issue three pad with leather band; hollow core rivets; later non-factory included M-31 variants.

Maker: Edelstahl and Thalestahl
Nomenclature: "M-34 Medium Duty"; "SD"; "SS-VT parade"
Years of manufacture: 1934-1938?
Shell characteristics: Shallow, dipped visor; embossed air vents; light blue primer; black enamel rivets; issued in field grey and black.
Liner: Special M-34 variant of the M-31 with cork rivet bushings, pressed paper liner band, and quick-release roller-buckle chinstrap.

Maker: Unknown - various
Nomenclature: "Civic"/"Police"/"Square Dip"
Years of manufacture: 1934-1945
Shell characteristics: Lightweight aluminum, often with square dip visor. Issued in black. Often feature "salt shaker" air vents.
Liner: Same as above, with addition of leather dome pad. Often had leather lugs for fire veil.

Maker: ET; Q, others*
Nomenclature: "German M-16-M-18"
Years of manufacture: 1916-1918
Shell characteristics: Chromium steel; stepped air vent lug for size up to 64; non-stepped lug for 66 and above. Manufactured in field grey; often re-issued in black top coating.
Liner: Three pad suspension with leather or steel liner band.

Maker: EB; others*
Nomenclature: Austrian M-16
Years of manufacture: 1916-1918
Shell characteristics: Shallow visor, non-stepped airvent lugs; commonly asymmetrical in shape; issued in brown in World War I and early SS users often refinished them black over brown.
Liner: Steel liner band with 3-pad suspension. Chinstraps often were cotton web with roller buckle and grommeted eyelets; or leather with roller buckle

Maker: Unknown - various
Nomenclature: "Droop-bill"
Years of manufacture: 1934-1936

Shell characteristics: Resembles Austrian M-16 with pronounced dip in visor.
Liner: Identical to the SS-RZM helmet. May have been same maker.

Maker: Unknown - various
Nomenclature: "Himmler helmet"
Years of manufacture: 1935-1937?
Shell characteristics: Resembles Austrian M-16 but with round dome and four liner rivet apertures. Finishes were black and field grey.
Liner: Pressed paper or leather liner band; 3-pad leather tongues stitched to band.

Maker: Erel, others
Nomenclature: Vulkanfiber or "fiberhelm"
Years of manufacture: 1935-1939?
Shell characteristics: Shells resembled the M-35 with a sloping visor and screen air vents. Many have a steel edge. Finishes include black and field grey.
Liner: Leather, reminiscent of civic style; pickelhaube style

General officers and security forces favored the lightweight helmets. The general rule of insignia use in 1934 can be characterized by two themes; hand-painted and non-standard. There are many factors governing what decorations and finishes the SS used. Some had to do with individual leaders' tastes. For example, period photographs show Reinhard Heydrich in Vulkanfiber, aluminum parade-style and RZM-style shells. Himmler on the other hand wears the round-domed civic style "Himmler helmet" almost exclusively in early 1930s photographs. This suggests SS leaders had much leeway in what they could wear as "acceptable." Conversely, SS units seemed to have, by 1934, adopted fairly standardized-and possibly symbolically-unique emblems for their helmets. In many photographs that can be attributed to the year 1934, the emblems on some of the units' helmets appear hand-applied. This is especially true of the LAH, SS-TV, and SS-VT units. Why was this evidently the practice? Perhaps by virtue of there being no organization-wide set of insignia, units took to using their own internal artists to decorate their helmets—perhaps this was the most efficient way to meet parade and other dress deadlines given the lack of a mature production system at that time. Mollo specifically states that some of these insignia, especially the circular ones, were manufactured (transfers). While several helmets bearing examples of transfer-type circular insignia have surfaced in the last decade or so, none of these proved to be original. One of the best-known surviving sets of (original) circular insignia is a hand-painted set. Also, a famous 1934 portrait of Sepp Dietrich shows him wearing what are undoubtedly hand-painted insignia on his black helmet. Many other period photos show hand-painted LAH and VT insignia.

A combination of natural wear-out and reconditioning of helmets, and ultimately, the general order of 1941 to turn in all transitional model helmets for melting down, undoubtedly wiped out the vast majority of the early SS helmets. Those that escaped the Straubing melt down and exist in collections today exhibit multiple layers of paint and various SS decorations within the layers. As for LAH and SS-VT insignia, until more original samples of helmets bearing these insignia can be identified, we are left with very, very few of these extremely rare decorated helmets.

One important insignia appearance in the 1934 period is the beginning of the use of "Fat Runes" decals by the LAH. In Leni Riefenstahl's film *Triumph of the Will*, there are SS-VT guards with hand painted circular insignia standing guard along the parade routes, while LAH bodyguard escorts are seen wearing SS-RZM pattern helmets with fat runes shields on them. This is an important clue that places these two types of insignia into the exact time frame of early fall 1934.

The list below serves as a guide to main units, their decorations and finishes:

Unit -1934	**Decoration**	**Finish**	**Camouflage**
Leibstandarte	LAH pattern shields	Black; black over field grey	None
SS-Totenkopfverbände	Skulls; Circular	Black; black over field grey	None
SS-Verfügungstruppe	Circular	Black; black over field grey	None
SS-Bereitsschaft Einheiten	Circular	Black; black over field grey	None
Allgemeine-SS	None	Black; black over field grey	None
Other (Reichsführung-SS; SS leaders)	None; combination	Black; black over field grey	None

1935 saw not only the most varieties of helmet styles and decorations, but also the eventual consolidation of decorations to one main type. The SS had grown into a huge political arm, as well as a fledgling field army. Because of their growth in numbers and mission diversity, the SS began to display a number of new decorations and finishes. These include the "Austrian" style decals in combination with the earth-brown finish; and the "fat runes" decal with the black and field grey combinations. In all, there were at least four decorations used by the SS in 1935. Finally, the end of the year saw the Reichsführer-SS establish the "standard pattern" runes and party shield combination for use by all the SS. The firm C.A. Pocher in Nuremberg produced this set of decals.

In terms of finishes and general helmet characteristics, the same basic patterns in use during 1934 carried over into 1935 with no new additions. The following table describes the helmet-user situation in 1935:

Unit -1935	Decoration	Finish	Camouflage
Leibstandarte	Fat Runes and 1st style/ Nat'l colors combination	Black; black over field grey	None
SS-Totenkopfverbände	Fat runes/Austrian pattern (?) and Circular	Black; black over field grey; earthbrown (?)	None
SS-Verfügungstruppe	Circular	Black; black over field grey	None
SS-Bereitsschaft Einheiten (Absorbed into the SS-VT)	Circular	Black	None
Allgemeine-SS	Combination of all	Black; black over field grey	None
Other (Reichsführung-SS; SS leaders)	Combination of all	Black; black over field grey	None

It is evident that there was quite a bit of variety within the many SS organizations regarding helmet styles, finishes and insignia.

User Units and Their Decorations: 1935

The Leibstandarte had gone from black to non-black (variations of grey) for their training uniforms and helmets, but retained black for parades. For a period of time early in 1935, the LAH are believed to have been close to adopting a grey helmet cover for their old style helmets. This idea may have been dropped because the cost of paint is markedly cheaper than canvas and leather; plus it wears longer and can be refreshed easily. There are quite a number of photographs of the LAH on parade in both black and non-black uniforms dating to 1935. In many photos, they are wearing the "Fat runes" helmet decal on both black and non-black helmets. There is footage from Leni Riefenstahl's *Triumph of the Will*, the epic coverage of the 1935 Party Day rally at Nuremberg, showing LAH soldiers wearing the fat runes. There are also a fair number of photographs showing the LAH still wearing the national colors tricolor shield in combination with the black on silver "CA Pocher"-style runes during the early to mid part of 1935. In some photos, their helmets are black, and their trainers' headgear is black as well. There is no official explanation for this evidently widespread practice-it may simply have been unit preference within the armed regiments.

The SS-VT adopted a similar field color concept for their helmets, but are seen wearing black helmets with non-black uniforms well into 1935 and possibly 1936. The SS-VT also is seen in these photographs to be wearing their circular insignia. Alas, the August 1935 order from the RFSS to standardize to the CA Pocher pattern decals included the SS-VT as well.

The SS-TV units are somewhat more diverse in their insignia during 1935.

OPPOSITE: Exceptionally fine portrait of an SS-Deutschland private wearing an Austrian M-16 painted black and decorated with the hallmark SS-VT circular runes and mobile swastika. *R. Noss*

In this period photograph of an SS-Berietsschaft formation, the men are wearing the circular insignia. This tradition existed in concentration camps to some extent, and most widely the SS-VT (which the Bereitsschaft units became). *R. Noss*

ABOVE: This grouping of SS Officers and NCO are wearing the Austrian M-16 shell with circular SS insignia. Judging by their uniforms, and insignia, this picture was likely taken in late 1934 or early 1935. Note the black helmet and field grey or earth grey uniform combination. *R. Rygaard*

LEFT: During the 1934 filming of *Triumph of the Will* Leni Riefenstahl was able to get excellent close-up footage of the SS-Verfugungstruppe who lined both sides of Hitler's route to the Nuremberg stadium. In this still, an SS-VT soldier wears an Austrian M-16 helmet with the mobile swastika insignia. Many of these appear to be hand-painted or stenciled rather than decals. In this same film, the LAH were wearing Fat Runes decals on their helmets. *D. Chapman*

OPPOSITE: Pictured side by side on this magazine cover are an SS-Unterscharführer in a German M-16 and an SS-Rottenführer in an Austrian M-16. Both have the final version of the standard issue insignia configuration – CA Pocher insignia (here visible the party shields). *D. Pajot*

Der Aufbau
Alleiniges amtliches Organ des Hauptamtes für Handwerk und Handel der NSDAP und der DAF für die RBG. Handel und „Das Deutsche Handwerk"
4. Jahrgang · Heft 19
Berlin, 2. Oktoberheft 1936
Einzelpreis 10 Pfennig
Leibstandarte Adolf Hitler
Zu unserem Bildbericht auf Seite 12/13

ABOVE: Another excellent display of the SS-VT helmet is evident on this cover photo from the period magazine, *Das Schwartze Korps*.

BELOW: Quite a few of the LAH soldiers in this photo wear Austrian M-16 helmets. Again, the size and positioning of the scalloped white on black shields seems to vary. *R. Rygard*

OPPOSITE: This excellent portrait of an LAH NCO shows a clear view of the helmet insignia referred to as the first style LAH runic shield. This example appears hand-painted. *D. Pajot*

ABOVE: This excellent rendering shows the basic characteristics of the LAH pattern insignia. *D. Pajot*

LEFT: Sepp Dietrich wears a German M-16 helmet finished in flat black with a narrow version of the scalloped national colors shield. *K. Hicks*

OPPOSITE: In this 1934 photo, a formation of the SS is wearing Hungarian, Austrian and German shells. They are refinished black, and display the national colors shield in what appears to be a transfer. While they appear to be LAH, there are no cufftitles visible. Note the man in the rear rank who wears a mobile swastika of the SS-VT on his helmet. *D. Chapman*

LEFT: This 1938 Anschluss period photograph shows an Austrian SS Lieutenant wearing an Austrian M-16 helmet with a hand painted runic shield on the left side. *D. Pajot*

This particular decal is the FW Quist variant of the first pattern runic shield. Quist or "Q-Pattern" runes were applied at the Quist factory, and differ from the CA Pocher runic shield in that the bottom plane of the runes extends beyond the lower break line of the shield by about 1mm. This subtle difference is common for all Q factory issue helmets. *C. Grayland*

Close-up view of what are commonly referred to as the "Austrian Pattern" runes. This decal is on an SD M-34 helmet, suggesting it is possibly another German variant, not Austrian. *J. Lichenstein*

"Austrian Pattern" runic shield. These decals have appeared on four known examples of Austrian M-16 shell with multiple layers of repaint. The time period of their use is uncertain, but is believed to be about 1935-1936. This decal is placed atop three layers of paint and two layers of decals, which account for the heavy cracking of the shield. *J. Tonzar*

Closeup of near-mint Q-Pattern runic shield on a rough texture finish. Note the 1mm extension of the runes beyond the break in the shield. The helmet is a Q-66 M-40, named to an SS soldier. *B. Smith*

CA Pocher pattern SS runic shield. Note that the runes are flush with the line of the break across the bottom portion of the shield. *J. Suprin*

"Mirror runic" shield, possibly of Austrian manufacture. The purpose of the backward runes is unknown. *B. Shea*

This cat is relaxing in a German M-16 helmet with a "Fat Runes" decal. *M. Horetsky*

Closeup of the "Fat Runes" decal. This decal is on an ET-64 M-35 helmet. These kinds of variant decals indicate that in the early prewar period, many companies were competing for contracts to produce items for the military. It is possible that this design was the forerunner to the "second pattern" or ET Pattern decal. *C. Jones*

CA Pocher pattern runic shield on a black finish, circa 1935. *K. Hicks*

CA Pocher pattern party shield on a black finish, circa 1935. *K. Hicks*

Mint example of an ET pattern runic shield on an M-35 helmet. Note how the bronze in the shield attains a greenish-brass tarnish with age. *D. Troiani*

ET pattern party shield, showing the thicker border characteristic of the ET shields. *D. Troiani*

ET pattern runic decal on an M-40 or M-42 shell. The ET runic shields on M-40 and 42 helmets have an aluminum-based shield, unlike their M-35 counterparts. They also appear to be of facedown manufacture because of their thinness. The bronze-based ET shields are thicker, suggesting face-up manufacture. This particular decal shows an unusually heavy amount of age cracking to the runic design. *B. Smith*

Original ET pattern runic decal still on the paper. *Private Collection*

RIGHT: Original ET pattern party shield still on the paper. *K. Bolton*

These LAH soldiers wear the standard CA Pocher party shield on their black transitional helmets. *R. Noss*

Closeup of the "Austrian" pattern large swastika shield. This decal was first identified in Goodapple and Weinand's Vol.2, pictured on a civic style helmet of unknown origin. This pattern swastika shield accompanies the special SS runic shield referred to as the "Austrian Pattern". The decals are very rare, but are considered to be standard issue rather than private purchase since they appear on a number of similar transitional helmets. Variations of many types existed among the prewar helmet decorations. *K. Hicks*

Unbordered large swastika shield on the backward or mirror rune SS transitional helmet. This pattern party shield is seen sometimes on early railway police and other civic type helmets. *B. Shea*

LEFT: This runic shield has been linked, through recent study, to the NS and SE helmet producers. It closely approximates the ET pattern decal but has a substantially copper-colored hue, earning it the name "champagne rune" (see Hicks, "SS Steel, Parade and Combat Helmets of Germany's Third Reich Elite" for reference to the champagne runic shield). *K. Hicks*

ABOVE: This party shield is the matching shield of the NS and SE pattern SS decal combination. *K. Hicks*

This runic shield is found on helmets manufactured by EF, after 1940. *K. Hicks*

Another NS pattern "champagne rune" decal. This narrower variety is common on NS and SE (hkp) M-40 and M-42 shells. *K. Hicks*

This NS pattern 'champagne rune' shield shows the narrowness and granularity of this decal pattern. *Private Collection*

CHAPTER TWO

THE TRANSITIONAL MILIEU

1936 still saw vast use of transitional and other non-standard helmets in use by the SS. The model 1935 (M-35) had been designed and accepted, with a limited number delivered to the LAH by mid-year. The new helmets were immediately given rigorous field-testing and while generally well received, had some flaws. Most notable of these was the tendency of the split rivets to shear under stress. What this meant was that with the slight up and down play in the area of the rivet aperture in the helmet and the liner (the former is a circular hole, while the latter is an oblong oval, the early brass rivets would break easily. This led to the manufacture of steel rivets for the helmets, but no other modifications to the aperture design. The second major problem with the M-35 again involved the liner. The early aluminum liner bands were of single-ply. The rivet holding the chinstrap loop in place proved to be another weak point. When a soldier strapped on a helmet snugly, then ran as fast as he could – as often takes place in military training – the chinstrap loop rivet would often tear right through the aluminum band or severely warp it. To avoid this, the liner bands were reinforced with an extra ply of the same gauge aluminum attached fore and aft of the chinstrap loop.

With the various problems solved on the technical side, the other big problem associated with the M-35 was supply. The manufacturers were hard-pressed to supply the huge demands of the Army and Luftwaffe, plus fulfill orders for nearly a quarter million helmets to be sold to the Nationalist Chinese. The bottom line was, with the SS at the bottom of the priority list for the new helmets, the supply of M-35s during 1936 was a small trickle (the SS was officially not a military organization at that time, and could not compete with the regular uniformed services for the latest military equipment items. This point will be raised again during the discussion of re-issuing of helmets which the SS did much more than the other services).

User units' finishes and decorations in 1936 continued largely as described in the table for 1935, with the exception of the insignia. It is believed that by early 1936 the SS were outfitting their units with the RFSS-approved CA Pocher Pattern SS decals. The LAH, SS-TV, SS-VT and Allgemeine-SS were refinishing and re-decorating their helmets, in black and field grey, and placing these new pattern insignia on top of the new finishes. To explore this practice a bit is to illuminate an extremely important aspect of SS helmets-the reissue and multiple uses phenomena.

Reissue and multiple uses of helmets by the SS were very common and resulted from a simple cause and effect relationship. The root cause was the lack of standardized supplies of headgear; dozens of makers scrambled to win contracts from the RZM to supply the SS with their products. The result was a fair variety of styles of helmets, but only a very limited number of useful combat shells for those who needed them most, the non-Allgemeine-SS. Commonly seen are helmets with several layers of finish and decals. SS units had parade, guard, and increasingly, field training as their primary missions. To keep their limited supply of helmets serviceable and looking smart, they were compelled to refinish and re-decorate them as often as necessary. Historically the position of the SS was subordinate to the Army, especially in the early days, in both equipment and manpower. Because of this, supplies were carefully controlled, especially field gear and helmets. In the 1930s, the most commonly reissued helmets were the old style ones. These include the World War I shells, commercial helmets such as the RZM and lightweight models, and others. (M-35 helmets from the later 1930s show the telltale signs of multiple issues, but the existing old stocks seem to have received the hardest use).

Pictured in this section are some examples of these helmets.

Unit -1936-1937: Leibstandarte
Shell: All previous and now, M-1935
Decoration: CA Pocher Runic and Party Shields
Finish: Black; black over field grey
Camouflage: None

Unit -1936-1937: SS-Totenkopfverbände
Shell: All previous and now, M-1935
Decoration: CA Pocher; Fat runes/"Austrian" pattern
Finish: Black; black over field grey; earthbrown (?)
Camouflage: None

Unit -1936-1937: SS-Verfügungstruppe
Shell: All previous and now, M-1935
Decoration: CA Pocher standardized

Finish: Black; black over field grey
Camouflage: None

Unit -1936-1937: Allgemeine-SS
Shell: All previous
Decoration: CA Pocher standardized
Finish: Black; black over field grey
Camouflage: None

Unit -1936-1937: Other (Reichsführung-SS; SS leaders)
Shell: All previous
Decoration: Combination of all
Finish: Black; black over field grey
Camouflage: None

This SS-VT troop wears a commercially produced droopbill SS-transitional. The price of these helmets was relatively inexpensive as is revealed in this archival note:
Article No. 93. SS Steel helmet (Reichswehr model), black (give head size) ..RM 7.50
SS-Befehlsblatt, Nummer 6, Munchen, den 15. Mai 1934, Nr. 18.
Photo courtesy Robert Noss

Right side view of an Allgemeine-SS or SS-VT helmet consisting of an Austrian M-16 shell, repainted black for use by the SS, with no insignia. The brown Austrian paint is visible in areas. This helmet configuration represents the earliest type of helmet used by the SS after the formal establishment of the Third Reich in 1933. *D. Pajot*

OPPOSITE: Interior view of the Austrian M-16 SS helmet, showing the standard horsehide liner, steel band, and wide leather chinstrap commonly found on these helmets. The inner dome retains its World War I Austrian brown finish. *D. Pajot*

LEFT: This SS sentry wears an M-16 helmet with the circular SS runic emblem on the right side. *R. Noss*

OPPOSITE

TOP: Black transitional helmet with the standard CA Pocher runic shield. The shell is the Austrian M-16. *K. Hicks*

BOTTOM: Left side of Austrian M-16 transitional helmet with standard CA Pocher party shield. *K. Hicks*

BELOW: The date stamp inside the liner provides the exact time period this helmet was re-furbished for issue to the SS. The liner itself is an early 1930s remake of the World War I style three-pad liner, but shows a much lower level of quality in materials and manufacture. Since the liner is a replacement, the dome of the helmet is painted completely black, instead of left as field grey, as with examples that did not have the liner replaced. *K. Hicks*

This formation of SS soldiers wears non-decorated field grey transitional helmets.

OPPOSITE
TOP: The national colors shield on this helmet is also hand-painted. *B. Hoffman*

BOTTOM: Probably one of the very few examples of an original LAH helmet with first pattern insignia. This helmet is a German M-16, refinished in satin black. The runes are hand-painted. *B. Hoffman*

OPPOSITE
TOP: Another transitional helmet with "Austrian" Pattern decals. This helmet also consists of an Austrian M-16 shell with field grey and black underneath the "earth grey" outer finish. The fact that these helmets saw service as black double decal transitionals suggests they were German, not Austrian, all along. Their exact unit of origin is impossible to determine until a marked example appears, or a photograph with undisputable supporting features turns up *B. Hoffman*

BOTTOM: Left side view, showing the characteristic "large swastika" shield. *B. Hoffman*

BELOW: This interior photo shows another variation of chinstrap, but the same three-pad and steel band liner construction. The field grey dome is visible, and there is earth grey paint overspray visible on the liner pads. *B. Hoffman*

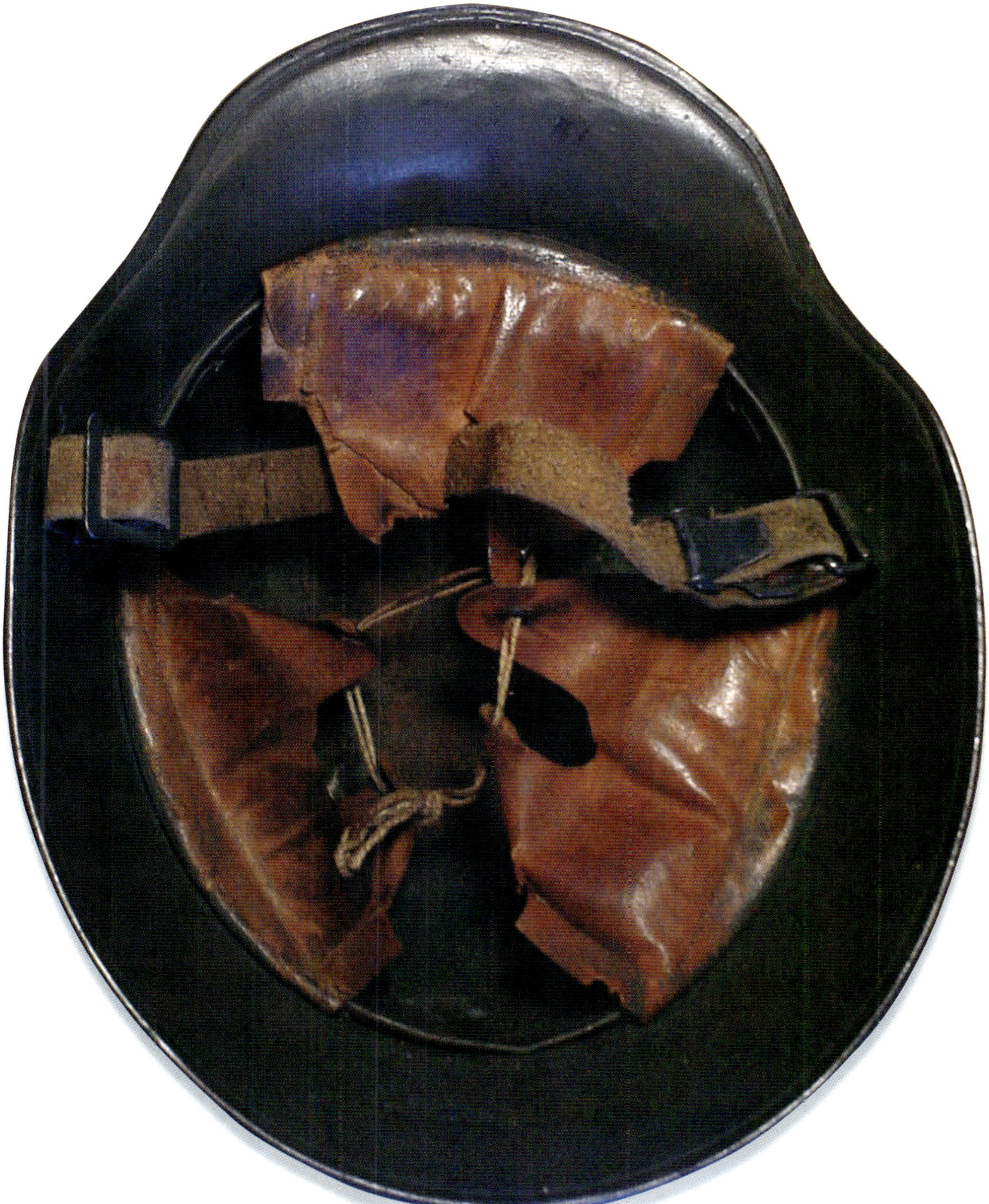

ABOVE: This example of the "Austrian" pattern transitional has its decals reversed, i.e., the runic shield on the left, and the party shield on the right. The helmet shell is again an Austrian M-16, with the same three layers of paint as the other helmets pictured. The outline of a party shield is clearly visible to the lower right of the runic shield. *K. Hicks*

BELOW: The right side view shows the large swastika shield. The cracks along the lower right correspond to the runic shield that is offset and underneath the paint. Cracking like this is common on repainted helmets. Also note the black and field grey finishes that are visible along the rim of the helmet. *K. Hicks*

ABOVE: This field grey SS-VT transitional helmet is in near-unissued condition. The runic shield is the CA Pocher pattern. *K. Hicks*

BELOW: The left side bears the CA Pocher party shield. *K. Hicks*

Minty interior of the German M-16 SS transitional helmet. The helmet was refurbished in the early 1930s, and the liner and chinstrap were added. *K. Hicks*

ABOVE: Austrian M-16 shell SS transitional helmet with Standard CA Pocher runic shield. *D. Pajot*

BELOW: Left side view of Austrian M-16 transitional with standard CA Pocher party shield. *D. Pajot*

Interior of M-16 helmet, showing three pad liner and field grey dome. Most domes were not painted because in refinishing, the liners were not removed. However, in examples where the entire liner system was replaced, the dome will sometimes be covered in the black refinish of the rest of the helmet. *D. Pajot*

ABOVE: Very unusual Black SS transitional helmet (German M-16 shell) with NS pattern champagne runic shield. *B. Hoffman*

BELOW: The left side reveals the exact NS pattern party shield. There is no official explanation for this decal occurence, although the helmet is clearly original. *B. Hoffman*

This formation of the SS-VT wears the circular insignia on their helmets.

ABOVE: Austrian M-16 SS transitional in field grey, with backward or "mirror" runic shield. There is no official explanation for this type of decal, which may have been a mis-print or meant for a double runic arrangement whereby both runes would face the front of the helmet. *B. Shea*

BELOW: Variant large swastika party shield on the Austrian M-16 field grey helmet. This helmet may have been painted black at one time. *B. Shea*

ABOVE: One of the rarest of all helmets is the black SS-VT transitional with circular insignia. This example bears hand-rendered insignia. *B. Hoffman*

BELOW: The left side of this helmet shows the mobile swastika design. There are theories that this exact helmet pattern, an Austrian M-16 style with pronounced rounded dome, is a commercially produced helmet. This style appears in many period photos of SS-VT troops. *B. Hoffman*

This portrait shows a transitional helmet in wear by an SS-TK troop. *D. Pajot*

CHAPTER THREE

COMMERCIAL, OR NON-COMBAT SHELL TRANSITIONALS

This early photo of a Niederbayern ceremony depicts transitional shells with large, hand-painted SS runes. *R. Noss*

This helmet is a lightweight version of the "Himmler" helmet, distinctive for its low scallop and rounded dome. This example features hand-painted runes, similar to the period photo. *D. Pajot*

RIGHT: The detail of these runes reveals excellent artistry and true ageing. *D. Pajot*

The left side of the helmet reveals a mobile swastika that is actually a decal. The finish is gloss black enamel and the year of manufacture is most likely late-1933/early-1934. *D. Pajot.*

LEFT: The close-up detail of this decal reveals a large amount of age crazing. This is a very desired characteristic, which cannot be duplicated. D. Pajot

This SS-TV officer uses his horse to model his "Himmler" helmet. Note the earth-grey M-36 uniform in conjunction with the black helmet. *D. Chapman*

This droop bill helmet shows several layers of paint, including an inner layer with hand painted runic and swastika insignia. The outer insignia is the CA Pocher design. The helmet, typical of the early SS, has seen much use and re-use. The party shield is the CA Pocher variety. *B. Hoffman*

The three-pad leather suspension system in the droop bill helmet is nearly the same as the SS RZM helmet. The Germans evidently experimented with a few basic shell and suspension designs, then produced ones that met the units' needs the best. Note the interior was left field grey, while the exterior was finished in black. *B. Hoffman*

ABOVE: Droop Bill helmet with Pocher pattern SS runes. This helmet was brought back from Berchtesgaden by a 101st Airborne Division veteran. *D. Pajot*

BELOW: Left side view of satin black droop bill helmet, with Pocher party shield. *D. Pajot*

This larger size droop bill helmet shows a layer of field grey or earth grey, painted atop a black finish. The shield is the CA Pocher make. The party shield is a matching CA Pocher decal. *D. Chapman*

The liner bears the property stamp "SD, RFSS, 1936." The earth grey over black finish is evident. The liner system is the typical leather three pad commercially produced system, using hollow core rivets to attach the liner to the helmet shell. The liner pads are constructed from one contiguous piece of leather. *D. Chapman*

ABOVE: This commercial parade helmet has an elongated visor and shallow dips. The runic shield is CA Pocher *B. Hoffman*

BELOW: Left side view of commercial parade helmet with Pocher pattern party shield. *B. Hoffman*

ABOVE: This droop bill helmet was purchased in a Berlin junk shop in 2004. Close scrutiny shows the runic shield was removed, evidently many years ago. *B. Hoffman*

BELOW: A smal portion of the party shield remains on the left side of this helmet. *B. Hoffman*

ABOVE: Fiberhelm with Fat Runes decal, probably from early-1935. The appearance of Fat Runes on any helmet is a rarity. *A. Barrows*

BELOW: Left side of fiberhelm, showing a heavily crackled party shield. This shield is the same as the basic Pocher design. *A. Barrows*

This civic style M-34 helmet is of lightweight aluminum and bears a Fat Runic shield on the left side. The purpose for this reversal is not clear, although it may have to do with SD or police type configuration of insignia. *D. Pajot*

RIGHT: The Fat Runes in detail show crazing and age. This is a very rare insignia by SS helmet standards. *D. Pajot*

The right side shows the typical CA Pocher party shield. *D. Pajot*

RIGHT: The party shield in close detail shows fine aging and no evidence of being placed atop other decals. *D. Pajot*

CHAPTER FOUR

THE SS-RZM HELMET

This photograph shows two or three commercial styles of SS helmet in use. In the early period of prewar SS, shortages of equipment were made up from existing supplies of commercially produced or contracted helmets. In this picture, Himmler, in front, wears the "Himmler" helmet, noted for its rounded dome and visible chinstrap rivets. Seated in the rear center is Reinhard Heydrich, wearing an SS-RZM helmet. The Nazi to his right also wears an RZM helmet. *K. Hicks*

ABOVE: This is the "classic" SS-RZM helmet in its original style of issue. There were clearly two finishes on issue RZM helmets; a distinct "gun metal" dark blue-grey, and black. Many were repainted dark blue and used by the Luftschutz late in the war. *D. Pajot*

BELOW: The SS-RZM helmet was produced under special contract exclusively for the SS. According to Baer, it was intended to replace the World War I style helmets as a true combat helmet. This did not take place. Note the closeness of the liner rivet to the air vent. This is a must-see on original RZM helmets. *D. Pajot*

The interior view of the SS-RZM helmet shows the leather suspension system. This was inherently flawed because heavy use warped and damaged it. For a combat helmet, this was not acceptable. *D. Pajot*

The RZM stamp, as well as the factory issue quick release chinstrap, are visible in this photo. *D. Pajot*

OPPOSITE
TOP: Possible use of at least one SS-RZM helmet in this early picture of an inspection of KL Sachsenhausen by Himmler, Eiche and others. *D. Chapman*

BOTTOM: In this 1934 photo, two SS officers wear RZM helmets in the early gunmetal finish, with possibly hand painted swastikas on the left side. *D. Chapman*

The SS stamping shows enough detail for the reader to discern the lightness of strike at the top and bottom of the circle. This is very much the expected look for this stamp on this helmet. Deeply struck stamps, or SS-RZM stamps on other than this exact model, are not original Third Reich. *D. Pajot*

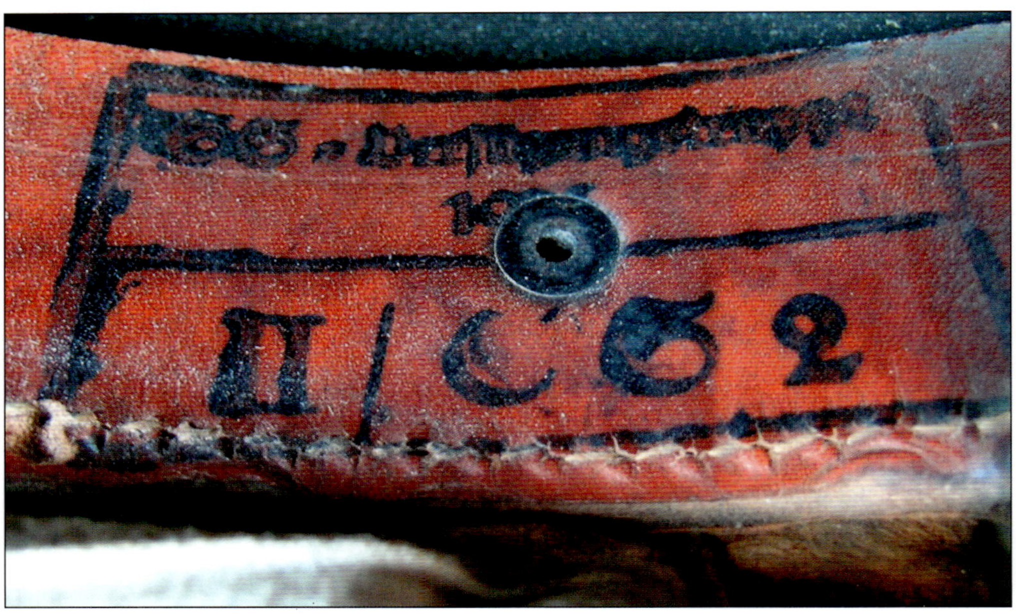

This property stamp reads: "SS Verfugungstruppe, 193?. II/SS 2," meaning the helmet was issued to a member of the 2nd battalion, Standarte "Germania." *D. Pajot*

ABOVE: This SS-RZM helmet displays the ultra-rare "fat runes" decal on top of a hand-brushed black finish. *B. Hoffman*

BELOW: The party shield appears to be the standard CA Pocher. The helmet is named to an SS-Unterscharführer in the LAH. *B. Hoffman*

This is a remarkably clear photograph of a young "Germania" SS-Scharführer wearing a brand new SS-RZM helmet. The decals, being unit issued, are the CA Pocher variety. *D. Pajot*

An SS-RZM helmet with standard CA Pocher runic shield. *D. Pajot*

RIGHT: The close detail of this decal reveals multiple layers of paint and multiple decals. Age, wear and heavy polishing (during the period) have made this decal a map of the desired features found in an original example. The SS seemed to be the one organization that made re-use of helmets a top priority in times of scarcity. *D. Pajot*

The left side of the helmet reveals the CA Pocher party shield. *D. Pajot*

LEFT: The detailed view of the party shield shows multiple decals, poor masking, and much wear. *D. Pajot*

The interior view of this helmet shows the standard leather suspension and chinstrap system unique to the SS-RZM and many other commercially produced helmets such as the Himmler helmet and droop bill helmet. There is evidence of multiple paint finishes on this example. *D. Pajot*

ABOVE: SS-RZM helmet with CA Pocher shields and factory black enamel finish. *B. Hoffman*

BELOW: The party shield is the standard CA Pocher. *B. Hoffman*

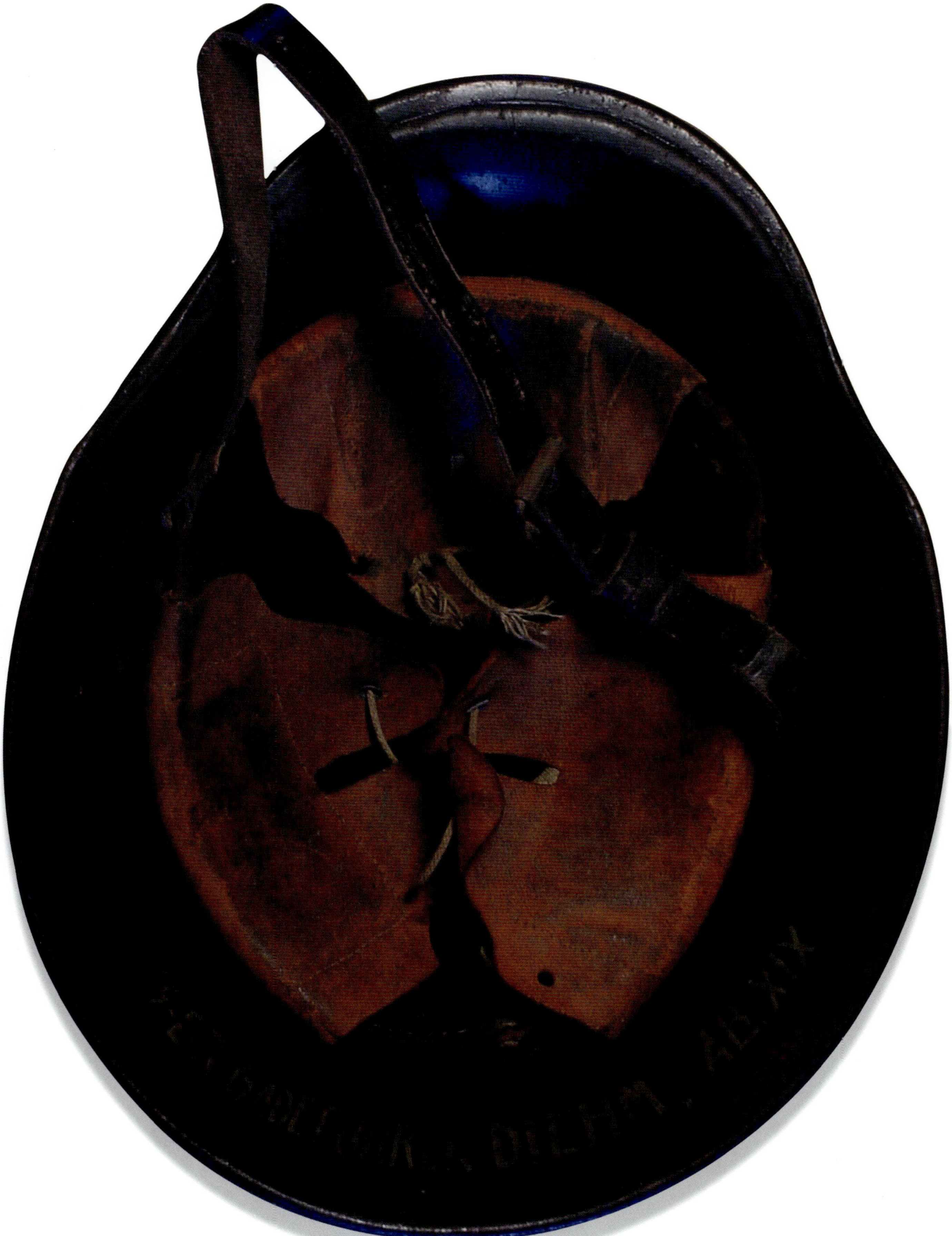

This helmet belonged to an SS-Brigadeführer Diehm, and bears an excellently rendered period inscription. *B. Hoffman*

CHAPTER FIVE

THE REARMAMENT AND THE MODEL 1935

1938 was the first year of full distribution of the M-35 helmet universally throughout the SS. All five major helmet manufacturers were in production for the Army and Navy procurement office, which supplied the SS along with the other services. The procurement office oversaw the acceptance and distribution of helmets to all the armed services. It must be kept in mind that Germany was undergoing a huge rearmament during this period, so the main effort of support for equipment was to the Army and the Luftwaffe. The Navy and then the SS received their issue of M-35 helmets. The procurement office had a representative at each helmet maker location; Abgenommen I through V. Roman numeral I usually appears in the dome of helmets manufactured at ET, II for SE, III for Q, IV for NS, and probably V for EF. These acceptance stamps were oval in shape, and were placed in the top inside of each helmet made. In many cases these stamps are not legible or visible, due to age, obliteration from patina, or not much ink being used on the stamp at the time. In practice, a helmet could not be considered fit for issue without this stamp, meaning the helmet had to be rejected or otherwise dealt with. Many original helmets that exist today exist have no dome stamp – it is just not visible today for one of the reasons given above. The postulation by some that this stamp was only placed on every hundred or so helmets as a symbolic acceptance is sheer nonsense – or lore, designed by modern counterfeiters to defraud collectors of their money.

Popular collector lore also once held that only ET and Q were producing helmets for the SS. Aside from the fact that there is no documentation supporting this, it is impracticable to make such an assertion about a military production and supply system. The fact is, in an expanding production environment such as that of Germany's re-armament period, all resources (in this case helmet makers) would be brought to bear to equip the using units. What really happened was a matter of sequencing of M-35 production coming on line with the manufacturers, as is described in Baer, and can be seen by reviewing the original examples in collections. ET and SE were the earliest and most prolific producers of the M-35 helmet. Both companies issued SS helmets with CA Pocher decals in the early period (except for ET's brief issue of "Fat Runes" decals), until they began local contract or otherwise using other patterns of decal ("ET" pattern and "champagne runes" respectively) later on. Q was the third highest manufacturer of M-35s, and NS, then EF. EF came into the production of M-35s so late that most observed examples exhibit the 1939 paint color scheme (see Hicks, *SS Steel*) or a rough texture finish; and are a less-often encountered shell in the M-35 configuration. So it was sequence and quantity of manufacture that accounts for the "appearance" of ET and Q being the most likely to have SS decals, but in fact all manufacturers produced for the SS on an as-ordered basis.

Another factor in helmet distribution was priority of support to using units. In the old Reichswehr and later the Wehrmacht, the procurement office was the entity responsible for filling orders for the services, which is why the Army and Luftwaffe consistently got priority over the SS (This condition was a result of the relatively low political power the SS held within the military hierarchy at that time). Sometimes seen are Army helmets, decorated as such at the factory, with SS decals placed atop the Army ones, due to the resulting shortage problems and other supply problems concerning the SS and their low stature in the early stages. What is almost never seen is a helmet bearing Army decals over the top of SS (this author has not observed such).

Regarding the M-35, three of the manufacturers were producing helmets with their own distinctive decals, which were contracted by local printers or otherwise available through the "parts system" (same as chinstraps, liners, liner rivets, etc). These makers were Q, ET, and NS. During the same time-and during the entire production period of the M-35 helmet, SE and EF are believed to have been using the National Pattern (CA Pocher) decals on the helmets they produced for the SS. SE may have adopted the NS pattern decal late in the M-35 period, as there are examples of these decals in use. The best evidence, an historical photograph, shows a clear example of an SE M-35 sporting the CA Pocher pattern runic shield. In the photograph, based on the decal's poor appearance, it is evident that the user may have applied the decal himself (see Hicks, *SS Steel*). The helmet is recognizable as an SE maker based on the upward sweep of the skirt toward the visor; the 'shallow visor' (This gives further support to the notion that the CA Pocher pattern decal was a normal supply item, as is strongly indicated in the vast amount of seemingly user or unit-level-applied

ABOVE: Excellent side shot of SS-VT-Ostubaf Otto Weidinger wearing an ET-66 M-35 helmet with CA Pocher runic shield. This picture was taken during Weidinger's early days in "Germania." *M. Yerger*

OPPOSITE: This portrait of an SS-Rottenführer provides a superb study of the ET helmet, including the ET Pattern runic shield. *R. Rygaard*

decals on the refinished black helmets of the prewar period. It also illustrates the universality of use of this pattern decal in light of Himmler's order of 1935).

M-35 helmets produced in 1938 were mainly decorated at the factory. The decals were applied by hand using an adhesive and two main types of varnish for topcoat; Damar Varnish and Ducolux lacquer. The surface was wiped clean, and a cardboard template with a shield shaped hole and cutout for the air vent were set against the helmet. The adhesive was then either brushed on, or sprayed, and allowed to set. The decals were trimmed and moistened by placing them atop saturated sponges. Once the card stock moistened, the decal's suspension compound would dissolve, allowing the decal to slip from the card onto the glued surface. A moist sponge would then be used to remove the excess suspension, as well as any air bubbles trapped between the glue and the decal. It is during this process that the problems occurred. Some original helmet examples show decals with wrinkles. This was a result of too much adhesive, which would pull the decal and distort it during drying. Otherwise, once the decal was applied and sufficiently dry, the topcoat of Damar or Ducolux would be brushed or sprayed on, again using a template to avoid excess lacquer around the decal. In almost every case, two things are noticeable on original examples: a fine, brown colored line of glue just to the outside of the border of the decal; and extending just beyond that, the outer edge of the lacquer coverage area. The latter accounts for the sometimes-visible lighter-colored line around the edge of the decal (the lacquer protects the portion of the paint it covers). Viewing other German helmets produced during this period is useful: they all had their decals applied the same way (if a factory-issue helmet).

Finishes to the early M-35 helmets distributed between 1936-1938 were of a uniform "pea green" matte finish for the non-Luftwaffe units. There is, as with most mass-produced items of

the time, a subtle difference between the different makers' colors of "pea green." ET is the standard, with a medium- tone green. NS has a noticeably olive cast to their finish; SE has a bluish-greyish version. Helmets produced by Quist tend to be slightly greyer than the others (all finishes are pictured in this chapter for comparison). Interestingly, 1938 was a year during which SS units were still refinishing them in black for use in parades. There is one example in this chapter of a 1938 Q-64 helmet that saw three different issues: factory production, use as a black helmet, and finally, reissued in matte grey for its final duty as headgear for a concentration camp guard at Dachau. The helmet has three layers of paint and decals, and is a living example of the resiliency of this fine model of helmet throughout its multiple issues.

Other shells in use in 1938 still include all of the previously mentioned patterns, although by this time the SS was well enough equipped that they were no longer relying much on the refurbishment and issue of World War I shells. An SS veteran who joined the SS-VT Regiment "Deutschland" in March 1938 told the author he was disappointed that his unit had done away with the World War I shells earlier that year. He said many of his comrades preferred the older style helmets with "horns" as having a more "classic appearance." He is pictured in this chapter wearing a brand new Q-68 shell M-35, re-issued in satin black for parade use. Other types of helmet shells, especially the lightweight ones, were still in use among the higher ranking officers. However, photographic evidence points to widespread use of the M-35 among field and company-grade officers by this time.

1939, the first year of war for Germany, was a monumental production year for the M-35 helmet. It was also a year in which there was a fundamental change in the color and texture of the factory issue paint finish of the M-35. At ET, SE and Q, the color changed from "pea green" to "apple green" with Q again being a grayer color. NS changed to a dark mud-colored finish. EF is unclear but may not have differed from the top three. There is one finish that appears late in 1939 that deserves mention as it appears on at least three SS and police M-35 helmets from that time. The finish is a waxy blue-green, and occurs on ET helmets with 1939 and 1940 liner band dates. In one example the paint scheme accompanies a 1940-dated reinforced liner band, a square and a round chinstrap bale (one of each!) and rough texture helmet rivet finishes. The dome stamp is also the later "Wehrmacht Beschaffungsamt Erfurt" one. This helmet, pictured in this volume, is ET-produced but sports the champagne "mock 2nd pattern" runic and party shield characteristic of NS. It and the other two clearly represent a late 1939-early 1940 transition period in M-35 helmets, manifested in the odd mixtures of early and late liner and suspension parts. Collectors who closely inspect their 1939 dated ET helmets will likely see chinstrap bales that have had their sharp corners purposefully rounded off like their M-40 counterparts that are described in the following chapter. They may also occasionally see the later style dome stamp in use. Some believe that it took up to six months in 1940 before the M-35 had been phased out, and the M-40 became fully integrated as standard issue. In this period, there were likely many "wear-out" declarations within the helmet factories, i.e., "use riveted bushings until supplies are exhausted," or "begin using rounded chinstrap bales in accordance with specification number x." This scenario flows more with a nation just entering a war and attempting to keep helmet design and production in step with the needs of the combat soldiers.

OPPOSITE: Near-mint condition SS M-35 helmet in satin black for parade use. The shell is a Q-66. Since all black M-35 helmets are technically repaints or reissues, the expected decal is the standard issue CA Pocher. This example has a very nice one of those. *D. Pajot*

ABOVE: Left side view of black M-35, showing some detail of the paint finish, which has dripped and run. "Allgemeine" helmets, as they are often called, frequently show poor quality of repainting. The Pocher pattern party shield is present. *D. Pajot*

This excellent quality photograph shows the fine detail of aging on this CA Pocher runic shield. The granular aluminum surface is visible, along with the other expected features: rounded rune tips, runes even with the break in the shield and finally, the base layer of acetate is visible in small areas where the metallic layer has flaked away. This decal has no lacquer coating, yet retains a highly silvery appearance and robust durability considering its age. When unlacquered, CA Pocher shields often are very bright. *D. Pajot*

The excellent detail of aging is visible on this closeup of the CA Pocher party shield. The white background is the acetate base layer. The red shield is the second layer and shows cracking. The swastika and border were the last layer printed. The black ink shows some age wrinkling, but not much overall wear because the helmet was well cared-for. This party shield is unlacquered. *D. Pajot*

The interior of the black M-35 shows its relative newness. This example was carefully painted on the inside, as the overspray has affected only the liner band and not the leather itself. *D. Pajot*

ABOVE: This Q-66 black SS M-35 helmet bears a runic shield that shows the effects of use and age. The CA Pocher decals were printed in three layers, starting with the sepia-white acetate, then a layer of aluminum pulver – as an ink – and finally the runes and border. To conserve on aluminum background material, the shield had a runic shaped cutout that was about 1mm narrower than the actual runic design. When printed, the runes overlapped the silver shield along their entire border by about 1mm. This is the reason that an inner line or border on the runes, such as the ones visible here, can be seen on many CA Pocher decals. This is a defining original characteristic of this decal pattern. BELOW: The CA Pocher party shield is featured on this Q-66 M-35. A note inside the helmet reads "This is to certify that I found the SS helmet (black) in an abandoned house in Luckenwalde Germany in 1944. Claude Smith." *K.Hicks*

ABOVE: Fine example of an ET 64 M-35 helmet that has been refinished in satin black and decorated with a CA Pocher runic shield. The helmet is in excellent condition and shows use commensurate with its utility as a parade helmet. *C. Jones*

BELOW: The party shield on this ET 64 Black helmet is the CA Pocher variety. *C. Jones*

RIGHT: SS-Mann Oswald K. Fernschild, of the "Deutschland" Regiment honor guard company, stands guard at the Temple of Honor, Munich, in 1938. He wears a Q-68 M-35 helmet. Black helmets were worn right up to the beginning of the war (late 1938-early 1939) and even into the early war years by some Allgemeine units. *O. Fernschild*

1.1. Steelhelmets:

All economical units of the SS-VT which have already been supplied with the field-grey steel helmet by the Procurement office of the army or navy, or their authorized representatives, must report the exact quantity to the Administrative office of the SS.

These steel helmets are to be worn exclusively with the grey service uniform. The black steel helmets still in possession of the units, after delivery of the field-grey helmets, are to be worn with the black service uniform (on guard duty, parades, etc.).

Again it is pointed out that the black steel helmets don't meet the requirements of field service. To avoid changes by mistake, it is strictly forbidden to repaint the black helmets field-grey, or the field-grey helmets with black paint. The unit officers are responsible for observing this order.

To retain a better fit of the black helmet, tests are presently being conducted with the new interior liner to be used for these helmets also.

Der Chef des SS-Hauptamtes, Verwaltungsanordnung Nr. 33, Munchen, den 29 September 1936.

This picture of a Totenkopf unit marching past is very unique in that it shows them wearing *black* M-35 helmets in combination with grey uniforms. This is an obviously early picture, but this combination of black helmets and grey uniforms has not been well documented in the past. *R. Noss*

ABOVE: Probably one of the first M-35s to be issued to the SS, this ET 64 helmet shows the early characteristics of brass liner retaining rivets and Fat Runes runic shield. Fat runes may have been the forerunner to the well-known ET (or "second") pattern decal, as the essential design involves an offset slice in the runes, rather than a straight slice. ET helmets were the first to be furnished to the SS throught the Army and Navy procurement office (Beschaffungsamt fur Heer und Marine). BELOW: Left side of ET 64 Fat Runes helmet. The party shield is a typical ET style, not a CA Pocher. RIGHT: This inscription means the helmet was issued to 2nd Company, SS-TV Regiment 4, "Ostmark." The single-ply liner band is dated 1937. The batch number is 3567. *C. Jones*

The ever-dapper Sepp Dietrich visits field training with Max Wunsche, wearing a regulation gas sheet, rubberized overcoat, and a brand new ET 64 M-35 helmet with Fat Runes shield.

ABOVE: Another example of the ET M-35 with early Fat Runes shield. This helmet is also an ET64 shell, and the batch number is 3574 (six digits off of the Chris Jones helmet). The liner band is also dated 1937, but the dome stamp appears to bear the date 1936. This helmet shows the earliest ET factory finish, which is a thinly applied paint similar in color to the later "apple green." While un-named, this helmet bears the history of having belonged to a member of the LAH staff at Berchtesgaden. BELOW: The left side of this ET64 Fat Runes helmet also bears the standard ET pattern party shield. *K. Hicks*

This Fat Rune M-35 is the third of four known examples. It shows heavy use, especially in being a reissue with an updated liner (two ply band). The decals were originally covered by the reissue paint finish, but the veteran uncovered the decals prior to selling the helmet to a friend of the author (in 1980). The helmet is named to "SS Grenadier Marzel Redhaber." The batch number is 3587. The range of these helmets is within 30 of each other. While the exact calculation of batch numbers is not clearly known, this is certainly a very narrow production run (a week?) of these particular helmets. *K. Hicks*

ABOVE: ET-64 M-35 factory issue helmet with CA Pocher runic shield. Prior to the approval of the ET pattern decals, commonly referred to as the "second pattern", ET used standard CA Pocher decals at the factory and at the unit on delivered but undecorated helmets. Many ET helmets that were part of the earliest shipments issued to the SS, arrived with no decals. This example has CA Pocher decals that were applied at the factory. BELOW: The party shield on this helmet is the ET Pattern. *J. Suprin*

ABOVE: This M-35 ET-64 helmet is another example of a pre-ET pattern decal configuration, consisting of first-issue CA Pocher decals. BELOW: Left side of M-35 helmet. *G. Aloisio*

ABOVE: Mint example of an ET M-35 helmet in stunning condition. This helmet bears the ET pattern runic shield applied at the factory. BELOW: The ET pattern party shield is evident on this example. OPPOSITE: Interior view. *D. Troiani*

ABOVE: This is a factory mint ET-66 M-35 with the pre-1939 "pea green" matte finish and ET pattern runic shield. *K. Hicks*

BELOW: Left side view of mint ET-66 helmet. The ET pattern party shield is evident. *K. Hicks*

Interior view of the ET-66 M-35 helmet. The date of manufacture is 1938, which is consistent with the reinforcements to the liner band (the early aluminum bands broke too easily and needed a reinforcing spar). The acceptance is clearly visible in the dome. It is believed that there is a close correlation between the office number (Roman numeral in center of stamp) and the helmet manufacturer. Many ET helmets have Roman numeral I, NS has Roman numeral IV, and so on. There is no evidence that dome markings differ whatsoever in Army or Navy M-35 helmets. These stamps were placed in the top of each helmet issued, not in every dozen or every hundred. The fact that many helmets do not appear to have them is due to the lack of sufficient ink on the stamp, or conditions of heavy wear. *K. Hicks*

ABOVE: Quist produced helmets that were distinctly different than ET helmets in profile. This Q-66 shows the classic Q features: shallow visor and scallop, straight and tall forehead, and a more squared skullcap than other varieties. The Q helmets produced before 1939 also have a slightly greyer cast to their pea-green finish than ET. BELOW: Left side of Q-66 helmet. *K. Hicks*

This closeup of the Q pattern runic shield shows the extension of the runes below the break in the shield. It also provides a good view of the characteristics of the glue and lacquer factory decal application. *K. Hicks*

BELOW: This SS-Kreigsberichter wears a Q-66 M-35 with Q pattern runic shield. *K. Hicks/Rex Reddick*

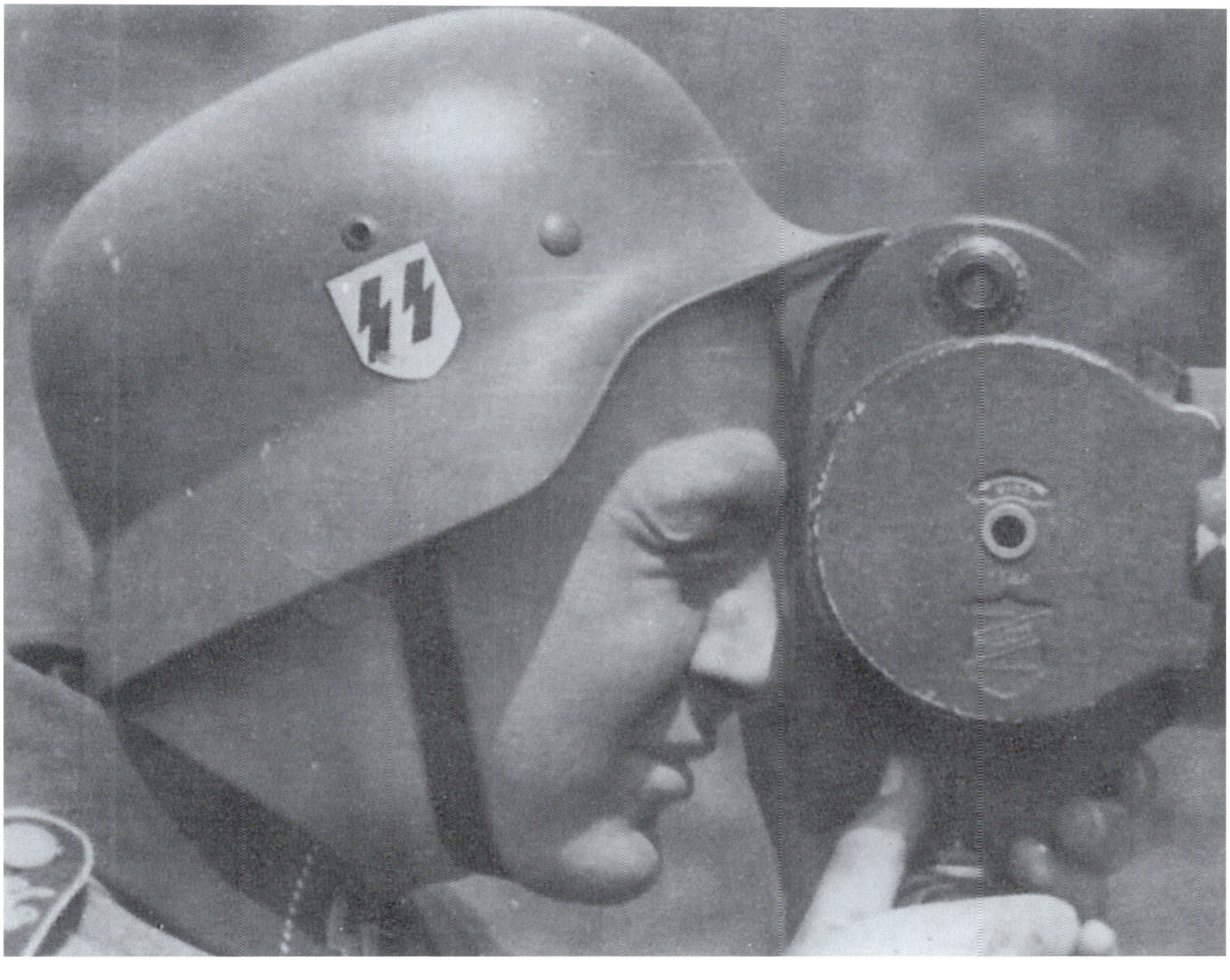

ABOVE: Moderately used example of an NS 64 M-35 helmet showing the dark grey-green finish typical of the NS manufacturer. Also present is the NS pattern 'champagne rune" runic shield. *K. Hicks*

BELOW: Left side shows the typical NS pattern party shield. The helmet is dated 1938 throughout. In the rear inner skirt is the inscription "Rottach, "Reich" e-SS h Abt." *K. Hicks*

ABOVE: A factory mint example of an NS M-35 helmet. This is another classic example of the NS pattern champagne rune. *J. Suprin*

BELOW: The left side view shows the NS party shield. The NS factory paint color is a dark grey-green matte finish. *J. Suprin*

ABOVE: This EF M-35 helmet is dated 1939 and bears the dark grey green finish of that maker. EF, like ET, Q and NS had their own distinctive brand of decal, but the EF M-35s are not believed to have borne it. This is because EF most likely did not develop their own pattern runic shield until sometime in 1940. This helmet was factory issued with non-CA Pocher SS decals and is re-decaled using the CA Pocher pattern runic and party shields. BELOW: Left side of EF 64 M-35 helmet showing the CA Pocher party shield.

ABOVE: Factory mint example of an EF M-35 production helmet with factory-applied CA Pocher decals. This helmet has a very matte finish that is characteristic of a late production M-35. EF is known to have come into the M-35 production scene later than the other makers. BELOW: The left side view of this EF helmet shows the factory CA Pocher party shield. *D. Troiani*

The interior view of this EF M-35 shows the standard M-35 liner configuration for 1938-1939. The finish is more the 1940 model configuration, suggesting this is indeed a standard M-35 "look" for the latecomer EF. *D. Troiani*

ABOVE: After the order came out in 1940 to remove party shields from helmets, many M-35 double decal helmets had the party shield scraped off by the owner. This Q-64 M-35 shows careful removal of the party shield. *K Hicks*

BELOW LEFT: The inscription reads "SS-Oberscharführer Rudolf Dimmel." Dimmel's file has been obtained from the National Archives. *K. Hicks*

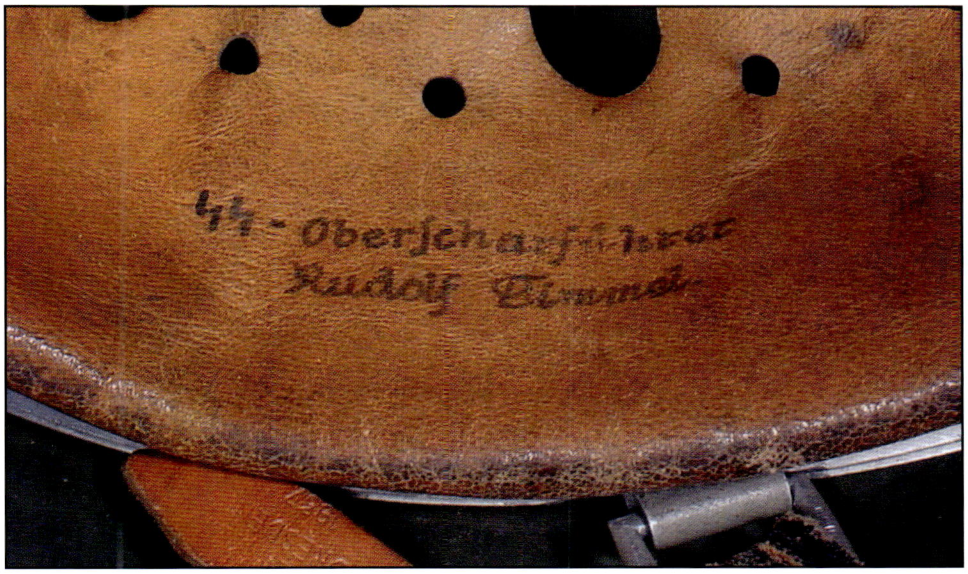

ABOVE RIGHT: Rudolf Dimmel was a member of the NSDAP in 1931, and a member of SS-Bereitsschaft "Oberbayern" in 1933. He went on to be an NCO instructor in II SS "D" and later in Dachau. He was last recorded as a member of V SS-Gebirgskorps in late 1944. He is shown here in his 1937 application for marriage from the "Race and Settlement Main Office" archive. *K Hicks*

ABOVE: SE also produced their own steel helmets for the SS, with the champagne rune as their primary decoration (as did NS). This runic shield pattern was used by SE until the end of decal application by the factory in late 1943. BELOW: The SE party shield was also the pattern used by NS. *J. Suprin*

ABOVE: Often, helmets are reissued and appear with multiple names, such as this Q66 M-35 helmet. Upon researching the names in this helmet, the owner discovered that both named SS troopers served at Auschwitz death camp. This is indeed a rare occurrence in helmet research. BELOW: The left side view shows the standard Q pattern party shield. *K. Niewiarowicz*

Detail of the first inscription: "Max Kolberg." *K. Niewiarowicz*

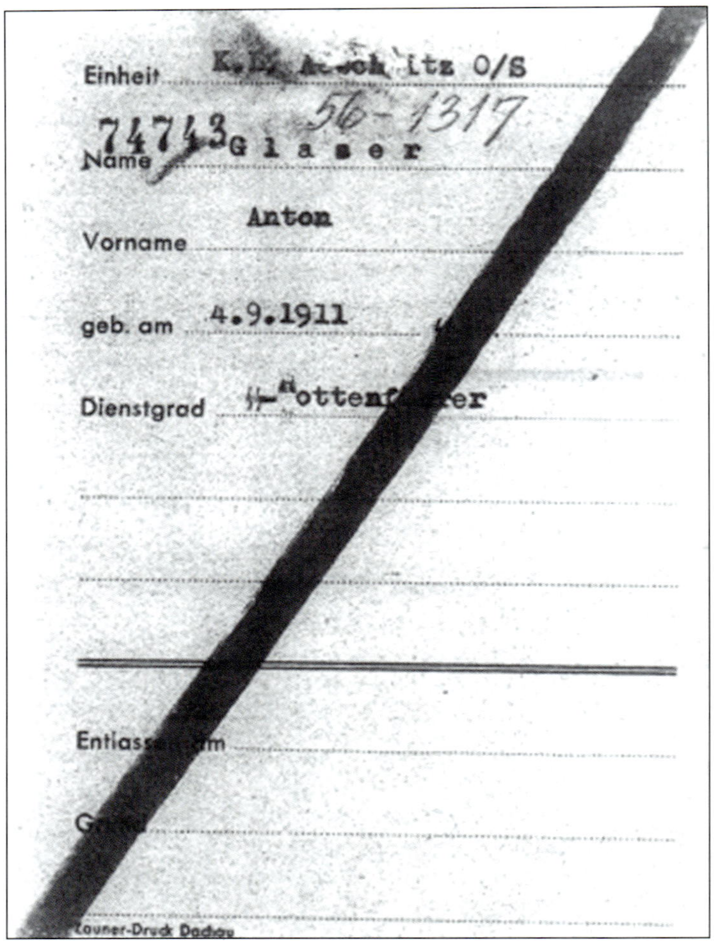

Einheit K.L. Auschwitz O/S
74743 56-1317
Name Glaser
Vorname Anton
geb. am 4.9.1911
Dienstgrad SS-Rottenführer
Entlassen am
Grund
Launer-Druck Dachau

The archival documentation regarding Glaser. SS helmets with a camp association are not common, but do exist, as the organization "SS" was in many ways affiliated with that dark part of their history. *K. Niewiarowicz*

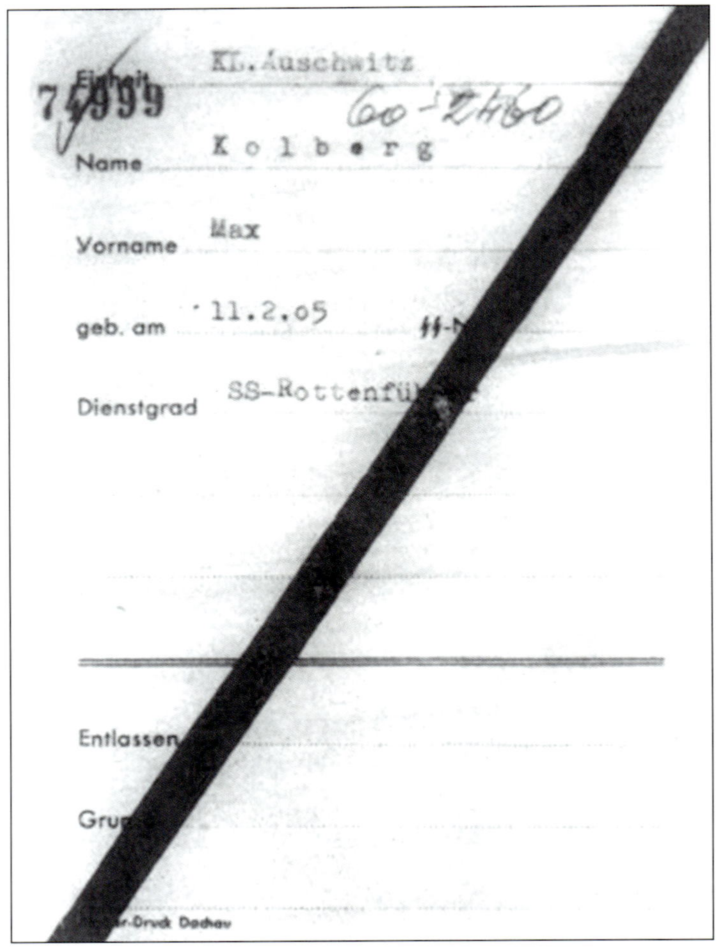

Einheit KL.Auschwitz
74999 60-2460
Name Kolberg
Vorname Max
geb. am 11.2.05
Dienstgrad SS-Rottenfü
Entlassen
Gru
Druck Dachau

Kolberg's archival documentation identifying him to the Auschwitz staff. *K. Niewiarowicz*

The second inscription: "Anton Glaser". *K. Niewiarowicz*

Two SS troopers from the SS-Heimwehr-Danzig wear their newly issued M-35 helmets. *R. Pygaard*

ABOVE: Nice example of a 1939-produced Q-64 M-35 SS helmet. Note the slate-grey finish and blue-green liner rivet color combination that was characteristic of Q, ET and EF during the later part of M-35 production. *K. Kithier*

BELOW: The party shield, as is the runic shield, are the expected Q pattern. This helmet is marked to a facility called "*Jugendschutz Möringen*", which was evidently a form of concentration camp. K. Kithier

CHAPTER SIX

THE WARTIME PRODUCTION HELMET: THE MODEL 1940

1940 was, as compared to the section on 1939, a year which still saw heavy use of the M-35 (during the first six to nine months of the year). Due to production limitations, the M-40 helmet was not actually fielded in great numbers until the latter quarter of calendar year 1940. Also, it is obvious from the hybrid nature of the 1940 dated M-35s (dull paint and ground chinstrap lugs; frequent use of rough texture rivets in combination with smooth finish) that there was a wear out period associated with the M-35. In militaries, old production continues to be used until supplies are no longer considered sufficient. There is usually a date designated when the item can no longer be used or worn. Concurrently, the new style would have been fielded during a one or two month overlap to ensure continuity of supply. It is likely that this was the situation that contributed to the hanging on of the M-35 long into 1940 until the M-40 could be universally supplied. It is worth note that many examples exist of M-35 helmets that were finished in the M-40 style rough texture paint and reissued as double and single decal helmets. This practice was especially evident in examples produced by the firm Quist in Esslingen am Neckar. Interestingly, in the Navy and SS, the reissues are multiple per helmet in many cases, suggesting the supply shortages to those services were much more acute in the early years.

The M-40 shell represented four major changes in German helmet manufacture: rough-texture non-glare finish, embossed instead of riveted ventilation holes, single decal usage and single ply liner bands made of zinc. There were also minor changes to the chinstrap buckles and studs, showing mostly mass-production oriented changes. Why the changes? It is important to remember that there was still a conservancy element at play in the German supply system. Raw materials were not in great quantity, thus subtle shortcuts such as embossed air vents likely produced substantial savings of time and cost in the manufacture process. Moreover, German armed forces were still in a dramatic phase of expansion, whereby cost and time saving steps would translate into more helmets being produced for the growing force. The third and not the least important factor affecting change in the M-40 helmet had to do with the direct experience of the 1939 and 1940 combat campaigns. The German soldiers recognized that smooth finish helmets required some type of camouflage or subdued surface in order not to stand out so much. Since the SS were already widely issued with excellent camouflage helmet covers by 1938-1939, the greatest pressure for improvement of the helmet finishes likely came from the Army, who did not have camouflage equipment at that early stage and needed something to make up for it-a change in finish. And since the very same supply and procurement system served all branches of the Wehrmacht, the SS had to live with what came out of the supply line as well-thus, rough texture finishes they could cover up anyway.

What collectors find is that during this period the double decal M-40 SS helmet existed in substantial numbers. The same does not appear to be true of the Army and Luftwaffe. This is quite enigmatic to consider-did the SS specifically order double decal helmets for their M-40 helmets? Were they then called on to conform to the armed forces standard of one decal? Evidently, as with the police helmets, the SS may have deliberately ordered double decal helmets well into the beginning of the period of changeover to single decal helmets. It is possible that the reason for clinging to the double decal tradition may have been because in combat and other tactical situations, the SS could solve the problem of camouflage by simply wearing their issue covers. Equally, the police had no tactical need for subdued helmets, and almost universally, the police M-40 and M-42 helmets retain double decals. SS double decal M-40 helmets are more the exception than the rule, but were clearly produced and supplied to units with two decals for a several month period in late 1940. Interestingly, the practice of double decals on SS helmets seems to have continued, albeit in very limited numbers, into 1943.

1941 is the first year that saw "exclusive" use of the M-40 helmet. There is not any major change in finish or decoration of helmets during this time. All units of the SS in combat are observed wearing the M-40 helmet. Other units, such as the SS-TV, the SD, or other SS-HA functionaries are observed wearing double decal helmets of many varieties.

This SS-Man wears an ET 66 M-40 helmet with ET pattern runic shield. *M. Davis*

The first M-40 helmets off the production line during the first half of 1940 were decorated with 2 decals. This practice was discontinued when the order came out for single decal usage, making a factory applied double decal M-40 one of the rarest of the Waffen-SS helmet varieties. This helmet is a Q-64 M-40 dated 1940 on the liner band and rivet ends. *K. Hicks*

RIGHT: The closeup detail of the runic shield on this M-40 double decal helmet reveals the Q pattern shield in use. *K. Hicks*

The left side view shows the distinctive Q pattern party shield in use. In the majority of original M-40 double decal SS helmets observed, the party shield tends to be the CA Pocher issue variety, implying that many M-40 double decal helmets were decorated in a non-factory setting. *K. Hicks*

LEFT: This close-up of the party shield shows the typical characteristics of the Quist party shield: narrower and redder field, with well-centered swastika. *K. Hicks*

OPPOSITE: Fine portrait of an SS NCO with an M-40 – probably ET helmet. *D. Pajot*

ABOVE: This helmet is a Q-66 M-40 double decal, showing a set of CA Pocher decals. This is unusual for a Q helmet, suggesting a period non-factory decoration. *C. Jones*

BELOW: The left side view reveals the non-factory applied CA Pocher party shield. *C. Jones*

ABOVE: This Q-64 M-40 double decal helmet was repainted and decorated with a CA Pocher runic shield. *P. Floyd*

BELOW: The left side of the helmet reveals the covered Q pattern party shield. *P. Floyd*

ABOVE: An example of an ET64 double decal M-40 helmet. The ET pattern decal is present. *J. Suprin*

BELOW: Left side shows a CA Pocher issue party shield, applied at an angle. *J. Suprin*

This Norwegian SS private wears an ET-66 M-1940 helmet that has sustained a significant hit from what appears to be a 12.7mm DSHK machinegun round. *M. Balestrand*

ABOVE: Factory mint example of an EF M-40 double decal helmet. Note the lighter grey "Schiefergrau" finish, and the distinctive EF pattern runic shield. *K. Hicks*

BELOW: The party shield on this EF-66 M-40 double decal helmet is a distinctive pattern that often appears on SE and EF police helmets. It gives a rare look at the party shield pattern that most probably corresponds to the EF runic shield when used as a pair on a helmet. *K. Hicks*

ABOVE: SE 64 M-40 finished in a dark charcoal colored Schiefergrau. This helmet displays the champagne rune in a factory-applied configuration. SE M-40 SS helmets generally are not abundant as SS helmets go; a double M-40 is even more unusual. *K. Hicks*

BELOW: The left side of the double decal SE M-40 reveals a non-factory applied party shield. This shield appears very similar to the CA Pocher type, but may be the variety found on SE and EF police helmets. *K. Hicks*

ABOVE: This helmet is an ET 64 M-40 single decal helmet, displaying a finely textured matte finish. The ET or second pattern runic shield is present on this helmet. *B. Smith*

BELOW: Beautiful example of a Q 66 factory issue M-40 helmet in near-mint condition. *B. Smith*

Interior view of the Q-66 M-40 in near-mint condition. This is an opportunity to view the M-40 liner system, with its single ply steel band and rounded wire chinstrap bales – mainstay characteristics of this liner system. *B. Smith*

This Q-66 M-40 helmet is a single decal, dated 1940. The rivets and liner band are all 1940 dated. *K. Hicks*

RIGHT: The runic shield close-up reveals the distinct Q factory issue decal with attendant characteristics. *K. Hicks*

ABOVE: Q-66 M-40 single decal helmet with expected Q Pattern runic shield, later "flat profile" liner pins and a 1942 dated liner band. Collectors often forget that the M-40 helmet was produced until the M-42 helmet came into full production. Many M-40 helmets are dated through 1942. *P. Sanders*

BELOW: Mint example of an ET-66 M-40 helmet. The ET pattern decal was factory applied and shows wrinkling during application. TOP LEFT PAGE 124: The close-up of this decal shows no wear, revealing how mint this helmet is. *M. Davis*

ABOVE AND RIGHT: EF 64 M-40 double decal helmet with removed party shield. This helmet is a rare occurrence in that it bears a "champagne rune," which is more characteristic of the NS maker. Interestingly, this helmet and the next example are identical and both produced by EF. These, as well as rare examples of EF M-42 helmets displaying ET or "champagne rune" decals, indicates that EF possibly sub-contracted from the same decal source used by ET and NS for a brief period. This explanation is logical, given conditions of supply shortage in wartime Germany. Still, the very clear majority of helmets produced by each maker (statistically about 92% of observed examples) bore the respective maker's own pattern of decals, as discussed in Hicks's "SS Steel". *J. Suprin*

OPPOSITE
BOTTOM: NS 64 M-40 with the dark green NS "Schiefergrau" finish. The runic shield is a champagne rune, expected on the NS maker. The liner band date is 1942. M-40 helmets were developed in 1940 and produced from mid-1940 until the M-42 production came on line (the model 42 was developed in 1942 and came into mass production by late 1942). Quist never stopped producing M-40 helmets, but did produce a limited number of M-42s. This NS helmet shows precisely the same manufacturer characteristics, including decal, as the chicken wire version in the chapter 8. *K. Hicks*

TOP RIGHT: Close-up detail of the runic shield shows the deep copper-gold color of this champagne rune. *K. Hicks*

ABOVE: EF M-40 single decal helmet with precisely the same characteristics as the previous helmet, illustrating the point of supply, distribution and use of insignia under varying conditions. *G. Currie*

LEFT: Close-up of the NS-Style runic shield on the EF M-40. *G.Currie*

OPPOSITE
Fine portrait of an LAH private wearing an ET-66 M-40 helmet. *M. Davis*

CHAPTER SEVEN

THE FINAL WARTIME PATTERN: THE MODEL 1942

In this chapter are shown many examples of the last pattern of combat helmet issued to the German forces and Waffen-SS. The Model 1942 was produced by all makers, although Quist is believed to have produced very few, opting instead to produce mainly the M-40 until near the end of the war.

The M-42 helmet differs from the M-40 in one key manner: the edge of the helmet was left un-crimped and ground lightly, eliminating several minutes of production time in this crucial period of the war. The non-crimping created the flared appearance of this famous design.

Another important change involving the M-42 is the introduction of German war production coding for two of the helmet factories, in the same manner as munitions and armaments. Thus, ET, the non-code maker's initials for Eisenhuttenwerke Thale, became the nondescript "ckl." SE became "hkp." But NS, Q and EF appear to have never been coded for the remainder of the war. Also, all makers' marks and size were moved from the left inside to the rear inside skirt of the helmet, right above the batch number. This change did not take place until June of 1943 according to German records. It must be kept in mind that very many helmets with a rear maker's stamping will still show a factory-applied decal, as the order to remove the latter did not come out until November of 1943, thus leaving six months of regulation production with decal, not to mention an undetermined overlap period of varying lengths per factory in compliance with the regulation. Keeping this important point in mind can help guide the collector in understanding why a "variation" to the regulation will sometimes occur.

OPPOSITE: Portrait of an SS private wearing an M-42 steel helmet with a first style runic shield. The general shape of the shield suggests CA Pocher manufacture. Helmet producers stopped supplying helmets with decals after November 1943, however, photographic evidence indicates the SS continued to wear them through the remainder of the war. The most commonly encountered runic shield used after late 1943 is the standard issue CA Pocher pattern. *R. Noss*

ABOVE: Factory mint example of an ET/CKL M-42 with second pattern runic shield. The shell size is a rare 70, with a size 62 liner. The inside skirt is stamped ET 70 (which is covered with factory paint) then re-struck in the rear of the skirt CKL 70, showing the transition from side to rear marking on shells. *K. Hicks* The eventual fate of factory issue and much extant insignia on SS helmets is laid out in this archival document:

402. Siegrunes on the Helmet

For the remainder of the war to come, the siegrunes as an insignia on the helmet is discontinued.

V.Bl.d.W.-SS, Nr. 21, 1 November 1943, Ziff. 402.

This M-42 helmet was manufactured by EF, and bears the telltale EF pattern runic shield and slate-grey "Schiefergrau" finish. *C. Grayland*

RIGHT: This close-up of the EF runic shield shows its slight similarities to CA Pocher, giving it a "first pattern" quality. There are in fact many differences, chief among them the high angle at which the runes are "sliced." This decal pattern is one of three possible found on the EF helmet; the other two being the narrow NS style rune and the ET style rune (chapters 6 and 8, respectively). *C. Grayland*

OPPOSITE
BOTTOM: Another CKL 70 helmet, in used condition, showing the classic ET pattern runic shield. *S. Kamisada*

The interior of the EF M-42 shows the standard suspension and chinstrap system characteristic of all M-40 and M-42 helmets. *C. Grayland*

ABOVE: Another example of the ET or CKL M-42, in mint condition and of an enormous head size. *B. Smith*

BELOW: This size 64 M-42 also shows the classic ET runic shield. Many M-42 helmets bear crookedly applied decals, attesting to the conditions of quality control and disruption during the mid and later stages of the war. *M. Davis*

Interior of the size 64 M-42 by CKL. *M. Davis*

Excellent example of an hkp 64 M-42, showing the conical vent hole shape characteristic of this helmet maker. *J. Suprin*

The detail of the hkp runic shield shows the narrow version of the champagne runic shield and its distinct copper hue. *J. Suprin*

Another view of the narrower champagne rune. *Private Collection*

ABOVE: Another hkp M-42 helmet, showing the distinctive conical vent hole and the wider style champagne rune, commonly found on NS M-35 helmets. *Private Collection*

BELOW: NS 66 M-42 helmet with the narrow champagne runic shield. The finish is a later war fine texture Schiefergrau. *K. Hicks*

ABOVE: CKL 64 M-42, showing the classic ET pattern runic shield. *P. Floyd*

RIGHT: This ET pattern runic shield is unique in that upon close examination, the soldier apparently removed the upper corner of the shield to create a scallop, much like that used by the "LAH." The helmet is named to and SS trooper named Walter Bohm, who served first in the LAH and later Totenkopf. *J. Gray*

Interior view of the CKL 64 M-42. *P. Floyd*

CHAPTER EIGHT

A STUDY OF CAMOUFLAGE PAINT FINISHES

Nice portrait of an LAH NCO wearing a camouflage painted M-35 helmet.

ABOVE: ET 62 M-35 helmet with typical second pattern runic shield. This helmet is tan and green sprayed camcuflage. The first owner, SS-Uscha Erich Malke, was a member of the LSSAH. The camouflage scheme is reminiscent of the Caucasus or Italy, both areas of operations for the Leibstandarte. *T. Smith*

BELOW: Left side of LAH camouflage helmet, showing the second pattern party shield commonly found on ET M-35 factory issue helmets. The shield appears to have been masked or carefully painted around. *T. Smith*

ABOVE: Q 66 M-40 helmet camouflaged in tan for tropical use. Underneath the SS runic shield is a bordered police shield. *D. May*

BELOW: Party shield remains exposed on the right side of this M-40 double decal helmet. The reason for the decal reversal is not clear, although it could be connected with police or einsatzkommando use. The inscription in the helmet reads "Heinecke, Hptm. d. Sch. Pol." (Captain of Defense Police). *D. May*

ABOVE: ET or CKL 62 M-42 Waffen-SS helmet sprayed in tan and green camouflage paint. The paint scheme is almost identical to the M-35 "Erich Malke" helmet. *R. Hamilton*

BELOW: M-42 Waffen-SS helmet, left side view of the camouflage paint scheme. This paint was spray applied. *R. Hamilton*

ABOVE: These two Waffen-SS volunteers appear to be participating in a training exercise or are posing for propaganda purposes. They wear helmets daubed with mud for camouflage purposes. The helmet in the foreground is an enormous EF M-42; the one in the background is a Q-66.

BELOW: NS 64 M-40 Waffen-SS helmet with factory finish and chicken wire covering. This helmet was taken from a newly deceased member of division "Hitlerjugend" by a US soldier during Operation Goodwood, the breakout of the Falaise-Argentan gap. During this engagement, "HJ" division was rendered combat ineffective. The helmet was sold by the veteran to the author in 1974. The veteran stated that straw was stuffed within the chicken wire, but he removed it so it would not detract from the helmet's appearance when he sold it! The finish and decal are the precise NS configuration. *G. Currie*

EF 64 M-42 with a decidedly autumn looking camouflage scheme. *S. Kamisada*

ET or CKL M-42 camouflage helmet, bearing a springtime color scheme of brick red and light green. *J. Faracco*

EF 66 M-42 helmet with a light tan or winter camouflage scheme. The camouflage finishes often wore much worse than the factory paint, due to their varying quality and thickness of application. *M. Davis*

RIGHT: This closeup details the EF factory issue first pattern runic shield, which has round corners and a higher angle of slice on the runes. *M. Davis*

Another Waffen-SS M-42 showing a second pattern runic shield and a dark camouflage scheme. The maker is EF, and the decal is a very unusual occurrence of an ET pattern decal on an EF helmet. This has been observed on only 5% of M-42 EF helmets, and suggests a brief period of time where the EF maker may have run short of supplies or lost their contractor due to disruptions in the supply lines (Germany was under constant bombardment in its industrial heartland, day and night, from 1943-1945). Observation of additional examples of EF M-40 and M-42 helmets with ET or Champagne runes also may indicate that EF used these shields for a substantial amount of time in the early stages of war production, then simply went over to the EF pattern rune once it was developed. *A. Barrows*

M-40 ET-62 Waffen-SS helmet with a light tan and green camouflage scheme. This color scheme may have been universal at a particular time, perhaps in the early stages of the Russian campaign. *P. Templeman*

RIGHT: This closeup of the ET pattern runic shield reveals the decal was painted around and left exposed by the wearer. *P. Templeman*

An exceptional example of an EF M-42 camouflage SS helmet with an original sniper's veil. *C. Jones*

OPPOSITE: Waffen-SS trooper wearing a chicken wire cover on his M-40 steel helmet. *M. Davis*

ABOVE: Q-66 M-40 single decal camouflage helmet, showing a light spray of red ochre color. *K. Niewiarowicz*

BELOW: Waffen-SS M-42 helmet with champagne rune and leaf pattern hand-painted camouflage motif. *S. Kamisada*

ABOVE: Waffen-SS camouflage ET-64 M-40 helmet. The color scheme simply consists of ordinance tan sprayed in swirls, leaving the original "Schiefergrau" exposed for the contrasting color. BELOW: Left side view, showing tan camouflage. *B. Maederer*

ABOVE: Excellent example of a spring camouflage scheme on a Q 66 M-40 single decal helmet. Note the full runic shield under the paint, and the marked rippling of the helmet shell from the stamping process. *S. Kamisada*

BELOW: The left side of the Q 66 spring camouflage helmet shows the green and ochre colors. *S. Kamisada*

SS soldier whitewashing helmets for his unit during winter operations.

ABOVE: ET 66 M-40 Waffen-SS helmet with whitewash for winter camouflage. *C. Jones*

BELOW: This Q 64 M-35 helmet retains a high percentage of whitewash camouflage and bears the Q pattern runic shield. The helmet is attributed to Knight's Cross holder Gustav Knittel. *F. O'Connor*

The front view of the Knittel helmet shows much whitewash remaining. *F. O'Connor*

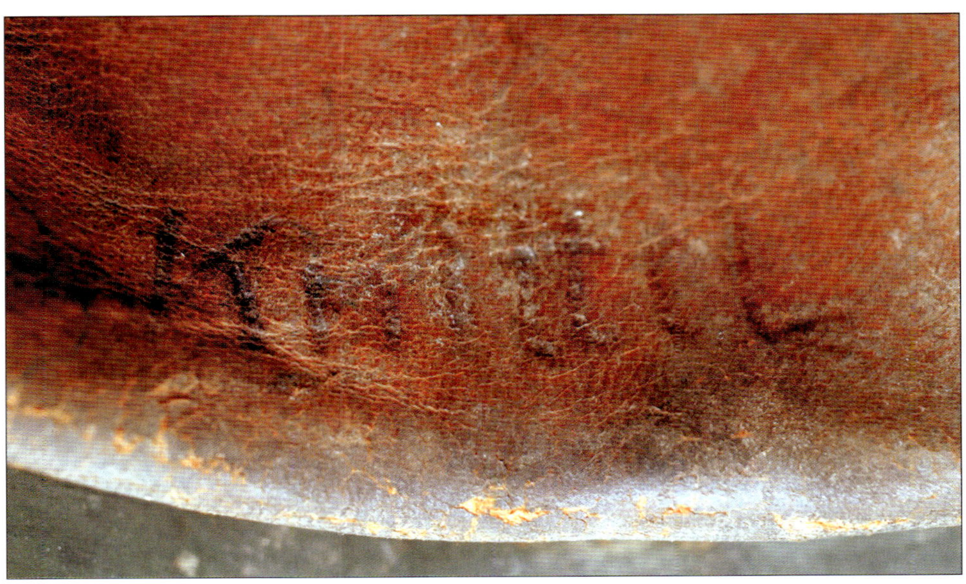

The inscription in the liner reads "G. Knittel." F. O'Connor

Gustav Knittel, upon award of the Knight's Cross to the Iron Cross. Knittel was sentenced to life in prison for his role in the Malmedy massacre on 16 December 1944. He was released from prison in 1953, worked as a car salesman for a period of time, and died of heart related problems in 1976. He commanded with distinction at almost all levels during his membership in the LAH. *D. Chapman*

ABOVE: Q-66 M-40 Waffen-SS helmet spray painted camouflage using two shades of green. The decal is just slightly visible under the paint. *C. Jones*

BELOW: ET-66 M-40 Waffen-SS helmet with a green and brown camouflage scheme. The decal was deliberately not covered in paint, remaining exposed. *C. Jones*

Waffen-SS M-40 Q-66 camouflage helmet with painted-around Q pattern runic shield. As a general note, original camouflage SS helmets seem to have been hastily painted, not elaborately or in multiple colors, as was the case with many Luftwaffe and some Army paint jobs. The chinstrap is a later pigskin variety. *J. Semen*

CHAPTER NINE

SICHERHEITSDIENST (SD)

This 1939 photo of three SD personnel shows two M-34 medium duty helmets and a commercially produced M-16-like transitional helmet being worn. The M-34 has come to be known as the standard SD helmet, although it is also known to have been used by the SS-Verfugungstruppe.

ABOVE: Rare example of a black factory issue M-34 helmet. The first pattern runic shield is from the firm CA Pœcher. *D. Pajot*

BELOW: There is no physical difference between this black M-34 and the field grey variety except the color. This helmet was most likely a parade helmet worn by a member of the Allgemeine-SS. *D. Pajot*

Field grey is the standard color of the classic SD helmet. In nearly all cases, the CA Pocher first pattern runic and party shields are found on these helmets. *P. Koelemann*

RIGHT: Closeup of the CA Pocher-produced runic shield. This decal shows no evidence of being glued or lacquered to the helmet, which is the expected configuration for the SD or VT helmet. *P. Koelemann*

The same applies to the party shield of SD/VT helmets. *P. Koelemann*

RIGHT: The CA Pocher party shield decorates the left side of this helmet. *P. Koelemann*

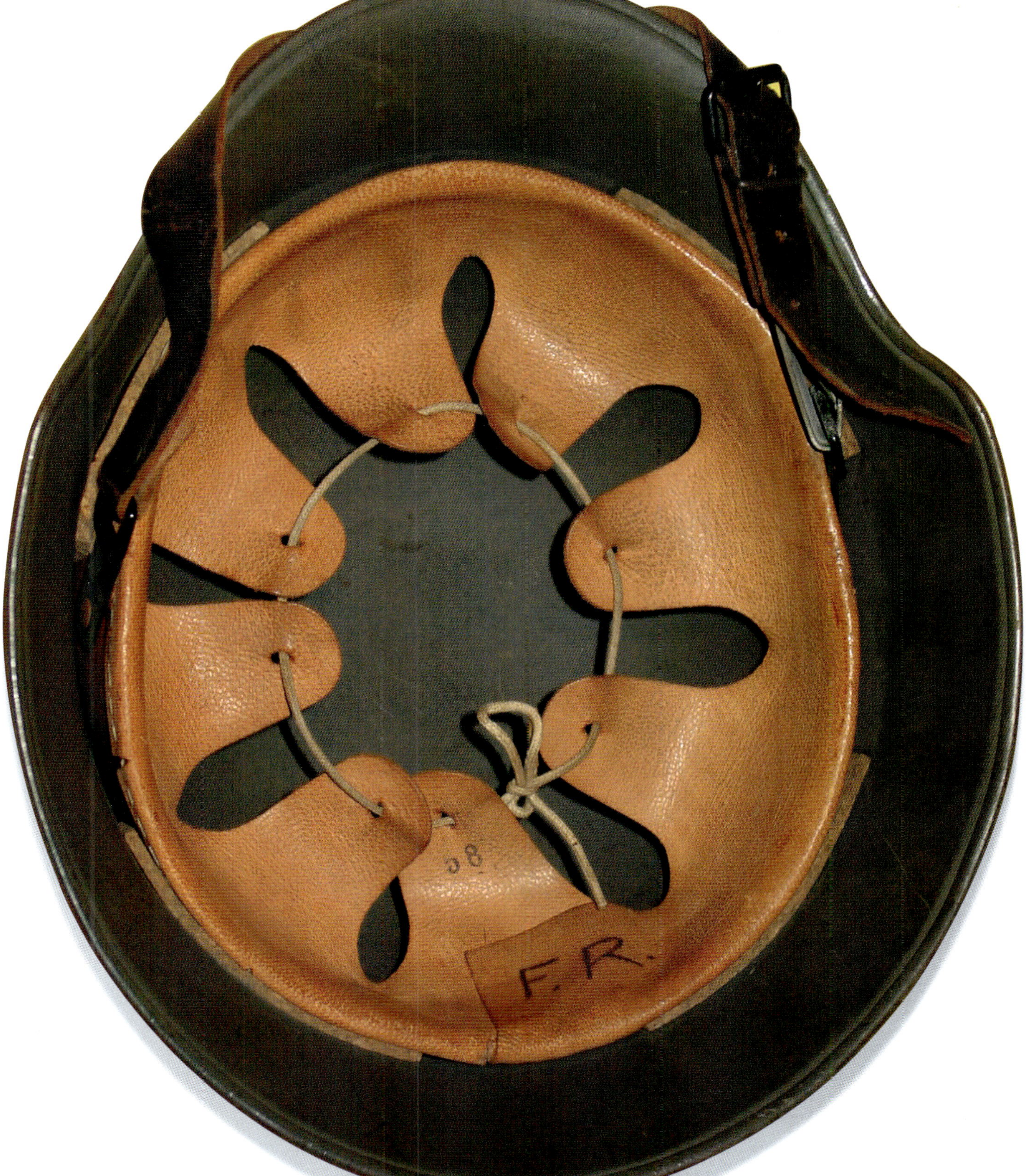

The suspension system of the M-34 medium duty helmet is of the civic variety, incorporating cork padding with cotter-pin retainer rivets. The liner bands are nearly always made of pressboard, a thick cardboard. The chinstrap is permanently affixed to the lugs by rivets; the lugs are attached on swivel rivets to the liner band. This is not an ideal configuration for combat use and would rarely be expected to appear in combat. *P. Koelemann*

ABOVE: This M-34 helmet displays the Nordic pattern runic shield. *Private Collection*

BELOW: This view shows a party shield that closely resembles the EF pattern party shield found on M-40 and M-42 combat shells. It is not the expected CA Pocher pattern party shield. *Private Collection*

This M-34 Medium Duty helmet is a single decal variety. *J. Lichenstine*

RIGHT: This runic shield is the rare "Austrian" version of the first pattern runes (*see chapter 2*).

ABOVE: Many SD helmets, for reasons unknown, are found in reverse decal configuration. *K. Hicks*

BELOW: The party shield is on the right side of this helmet. The shields are CA Pocher first pattern. Some SD helmets have two layers of decals, one in reverse configuration and one in standard. This helmet is factory issued in reverse. *K. Hicks*

Very unusual camouflaged M-34 SD helmet. The helmet is sprayed in brown and dark green stripes. *Collector's Guild*

RIGHT: In this closeup of the CA Pocher first pattern runic shield, the brown and green camouflage colors overlap the decal slightly. *Collector's Guild*

Left side view of camouflaged M-34, showing the strong camouflage colors. *Collector's Guild*

LEFT: This closeup of the party shield shows it was masked off rather than painted around. *Collector's Guild*

OPPOSITE
TOP: This photograph shows the same Estonian volunteers at an award ceremony – displaying an interesting array of helmets: the soldier shaking hands wears an M-34 medium duty helmet, while the men on each side of him show the reversed runic configuration on their helmets. The man at the far right of the photo wears an unmodified Army helmet.

BOTTOM: These Estonian volunteers wear a mixture of helmets: on the right, a double decal M-40 that is likely a converted police helmet with SS shields; and on the left, an M-34 lightweight police helmet.

CHAPTER TEN

VARIANT AND NON-MAINSTREAM HELMETS AND DECORATIONS

The impact that foreign units in the SS had on insignia used on uniforms and helmets cannot be underestimated. In the multitude of excellent reference books by Mollo, Angolia, Beaver, Lumsden, Yerger and others, there are excellent photographs and analysis of many foreign versions of insignia. In helmets alone, there are at least three identifiable variant versions of the SS runic shield. In configuration of the decals, there are two main methods – reverse decal and double runic. These will be explored below.

Double runic helmets are the first variation to be discussed in this section. Photographic evidence from the World War II period suggests these made their first appearance during the campaign in the Low Countries, May-June 1940 (see Hicks, *SS Steel* pp. 196-197). Other photographs from the same period provide excellent, clear examples of the practice of wearing two runes, as well as the bonus of the wearers' unit identification being Regiment "Germania." Why did the SS apparently widely use two runic shields in this period? One answer may be that the practice was related to the wearing of double runic collar patches by certain combat formations of the SS during a similar period and into 1941. The order to abandon two decals – in other words the party shield from SS helmets (all services' helmets) – came out in late 1940. It is possible that SS units in the field chose to interpret the order to cover up the party shield with paint their own way and used a runic shield instead. Some photographs predate this order, and are an enigma. If the purpose of two runes was to subdue the colorful party shield, certainly a bright and shiny SS runic shield is not the answer. Though still unanswered as to purpose, at least some light is now shed on the identity of at least one unit using double runes, "Germania."

Reverse-decal helmets, another commonly encountered variant helmet, existed within the SS formations as early as the double runic variety – from nearly the beginning of the war. The vast majority of these helmets are M-34 Medium Duty SS-VT or SD helmets (sometimes marked Edelstahl or Thalestahl). There is no known explanation for the reversal of decals, other than possible police or security affiliation, or mistakes. The latter is not feasible however, due to the relatively large number of existing examples of reversed and re-reversed M-34 helmets. Other SS helmet specimens display reversed decal configurations. These include an earth-brown transitional helmet, a "fat runes" reversed decal lightweight commercially produced helmet; a model 1935 helmet once issued to an Einsatzkommando Officer in the SD, as well as multiple examples of M-40 and M-42 combat police helmets converted to SS use. Other than named examples that indicate the owners' unit, there is not a hint – any official or semi-official explanation – for the reversal of decals on SS helmets. In this regard, they are in the same category as double runic helmets – rare and unusual. As for the M-42 variety, there is an explanation in a series of photographs of Ukrainian SS troops wearing this type of helmet. One example the author has observed actually had an authentic inscription in Cyrillic, the Soviet Ukrainian script of the war period. This is much more indicative of a deliberate or official policy for the specific use of that helmet. First, it is widely evident that many foreign volunteer formations in the SS wore their runic shield on the left instead of right side of their helmets. This was universal enough to have been issued as an order in one of Himmler's Verordnungsblatt der SS. Second, many foreign volunteer units, especially Ukrainian, were initially police formations. Many of these were anti-partisan in nature. Having been issued with police helmets, it is logical that when the Ukrainian forces were standardized into the SS, they simply added the runic shield atop the police shield. This configuration is the case in 100 percent of observed real examples of this helmet.

ABOVE: This M-35 is an ET 66 shell that has been reissued, camouflaged with dark green paint mixed with wood chips and decorated with an SS rune on each side. *K. Hicks*

BELOW: The left side of the double runic helmet reveals a CA Pocher pattern runic shield. These helmets are known to have been worn by the "Germania" regiment during the French Campaign, as well as by some foreign units in the SS. *K. Hicks*

Closeup of the right side runic shield showing that the decal was placed directly atop the heavily textured camouflage finish. *K. Hicks*

The left side runic shield is also a CA Pocher, which was placed atop the party shield. Through a nick in the right rune can be seen the arms of the swastika decal underneath. *K. Hicks*

LEFT: In this picture of men from the SS-VT Regiment "Germania," both are carrying double runic M-35 helmets. *R. Noss*

OPPOSITE: This young SS-Mann wears an M-35 double rune helmet. His old-style epaulets indicate this picture was taken in the early stages of the war, if not late pre-war.

ABOVE: This helmet is an ET70 M-35 with a size 63 liner. The name Heiss appears in the skirt of the helmet. It bears a 1938 date in the liner, a factory applied CA Pocher runic shield on the right, and a field applied CA Pocher shield on the left. *P. Sanders*

BELOW: The left side of the helmet reveals the CA Pocher runic shield. It is applied atop a removed party shield. The helmet finish shows patina from storage in a NY City attic since the end of World War II. *P. Sanders*

ABOVE: ET 66 M-40 with ET pattern runic shield. *A. Barrows*

BELOW: The left side shows a CA Pocher pattern runic shield applied in the field to make this helmet a double runic. *A. Barrows*

This ET M-40 double runic helmet features the classic ET pattern decal on the right side. *R. Kondor*

The left side features the expected CA Pocher pattern runic shield, applied in the field. *R. Kondor*

This helmet came from the original source complete with the first model camouflage cover used in conjunction with this helmet. This array shows an early usage of the double runic helmet by a German, non-foreign SS soldier. *R. Kondor*

ABOVE: This ET 66 M-40 helmet was part of a grouping from Norway. The helmet was field re-finished in a medium grey, and decorated with two variant SS runic shields. *C. Jones*

BELOW: The left side of this helmet shows the variant runic shield. *C. Jones*

In this propaganda photo, Dutch volunteers simulate firing their sniper weapons. Their reverse-rune helmets are most likely not double runic, but rather single decal helmets, as worn by many volunteers from the Low Countries. The Finns wore double runic helmets as foreign volunteers. *J. Borsarello*

Closeup of the Norwegian variant runic shield. This same pattern is featured on the "Schalburg" helmet in this chapter, as well as on the SD helmet in that respective chapter. The manufacturer of this decal is unknown, but definitely was designed and issued in Scandinavia, or produced in Germany and directed for issue in those countries. *C. Jones*

The same variant pattern runic shield appears on this side of the helmet. Compare this to the M-40 "Schalburg" decal – they are identical. *C. Jones*

ABOVE: The Finns also wore double runic helmets as part of their combat uniform. This mint EF-66 M-42, still in Finland, is from a Finnish SS veteran. Note the EF pattern runic shield. *Private Collection*

BELOW: The left side of the Finnish volunteer helmet shows the expected CA Pocher pattern shield. *Private Collection*

ABOVE: Another often-encountered variant helmet of the SS is the reversed decal variety. This example consists of an M-40 Q-64 police, with the addition of an SS rune over the police shield. The two distinct uses for this decal configuration were Sicherheitsdienst or foreign volunteer. The police shield is visible where the runic shield has flaked off. BELOW: The right side of this helmet reveals a party shield decal. *P. Floyd*

This M-42 helmet was a refitted and repainted (Luftwaffe) helmet that bears a variant, possibly foreign-made, decal. *P. Sanders*

RIGHT: This M-35 reissue helmet bears precisely the same variant decal as the previous M-42. The foreign context is most probable in the origin of this runic shield. *B. Shea*

Excellent example of a very rare Danish "Schalburg" helmet, named and documented to a Danish volunteer. This helmet consists of a Q-66 M-40 double decal helmet, painted over in gloss black, and given a Norwegian pattern runic shield. *D. Pajot*

RIGHT: Closeup of the Danish pattern decal. Note the similarity to the CA Pocher design. This rare variant is known on three original helmets. *D. Pajot*

NS-62 M-35 helmet painted medium grey green and decorated with an Italian fascist eagle. This helmet is an outstanding example of what the SS volunteers frequently did: decorated their helmets in a haphazard manner. Period photos of Italian SS troops invariably reveal many different styles of helmet decorations. *K. Hicks*

RIGHT: The national emblem on this helmet shows a high degree of artistic ability in its application. The best assessment is this helmet was an individual's attempt at decorating his own helmet – not indicative of any official or widely used design. The helmet is from the World War II period and is most likely not U.S. GI art. *K. Hicks*

The left side of this Italian helmet bears a stylized SS runic shield. *K. Hicks*

RIGHT: Closeup of the runic shield reveals the telltale signs of age and originality. Note the field grey factory finish underneath the medium grey. This helmet was likely an Army issue helmet prior to the SS configuration. *K. Hicks*

ABOVE: This Schalburg helmet is a beautiful example that once belonged to Ludwig Baer. The mobile swastika is stenciled and the satin black finish is sprayed on the helmet. *D. Pajot*

OPPOSITE: Italian SS volunteer in relatively conventional dress. The helmet is a single decal M-40. *R. Noss*

CHAPTER ELEVEN

THE SS-"DEUTSCHLAND" PORTFOLIO

A grouping of about one hundred glass-plate negatives, containing portraits of enlisted members of SS Regiment "Deutschland" recently surfaced in France. These photographs were taken in 1940 while the regiment was lagered at Preigney, Haute Saone, France. The candid shots of these men show them relaxed and wearing no combat gear. Their tunics display a mixture of new pattern runic collar insignia, with older SS-VT style sleeve eagles. The grouping of photographs is indeed an incredible "time capsule" of helmets and uniforms, providing a rare opportunity for precise analysis of styles and insignia of the mid to late 1940 Waffen-SS.

The portraits selected for this section show four of the five helmet manufacturers' product in use, with the exception of NS, which does not appear among the 30 photographs showing helmets in wear. Present are a majority of ET with both CA Pocher and ET runic insignia; Q, with standard Q pattern runic and party shields; SE with both CA Pocher and champagne runic insignia; and EF, with field-applied CA Pocher insignia. Moreover, the majority of the helmets in use in these portraits are the M-35 pattern, with only several examples of M-40 helmets shown.

The captions accompanying each photograph provide more information about these excellent and rare portraits. Special thanks go to Mr. Marc Negrault, who has kindly lent prints right from the original glass plate negatives.

This portrait shows an ET 66 M-35 helmet with CA Pocher runic insignia. The helmet bears the pre-1939 finish and in combination with the Pocher runes would be late 1937 or early 1938 production. Negrault/Hicks

This SS-Mann wears a classic ET 66 or 68 1938 production with ET pattern runic shield. Negrault/Hicks

This SS-Sturmmann wears a classic Q66 M-35 with Q pattern runic shield and possible 1939 finish (denoted by the darkness of the color). Negrault/Hicks

This SS-Mann wears an EF 64 M-35 with replaced CA Pocher insignia. Close inspection reveals the runic shield was placed over another insignia that was scraped off. Negrault/Hicks

Excellent portrait of an ET 64 M-35 helmet with CA Pocher runic shield. Negrault/Hicks

This SS-Mann wears a Q 66 M-35 with party shield in use. This clearly shows that the SS were wearing double decal helmets well into 1940, despite orders and other documentation suggesting the wearing of double decals was abolished earlier that year. There is a lot to say about this practice, but the point clearly is that it took time for policies and orders to be universally adhered-to among the field formations Negrault/Hicks

This SS-Rottenführer wears a Q 64 M-35 with party shield intact. The finish is reminiscent of the darker 1939 color. He also wears the tall head SS-VT sleeve eagle. Negrault/Hicks

This SS-Mann wears a very interesting M-35 Q 66 M-40 style shell, with M-40 factory finish but M-35 air vent bushings. The Q pattern decal shows signs of flaking, characteristic of the adherence problems on the rough texture finish. The helmet was no doubt a hybrid of parts from the last of the M-35 parts utilization, and was probably a double decal at time of issue. An example of this type of production appears in this book, attributed to T. Spier. Negrault/Hicks

The ET 66 M-40 helmet is shown in this excellent portrait. Negrault/Hicks

This SS-Sturmmann wears what appears to be an SE 64 M-35. The second pattern runic shield is likely the "champagne rune." (the size of the air-vent hole and shape of the scallop are reminiscent of the SE design). Negrault/Hicks